Fatigue and Driving

Driver Impairment, Driver Fatigue and Driving Simulation

Edited by

Laurence Hartley
Institute of Research in Safety and Transport
Division of Psychology
Murdoch University
Western Australia

Taylor & Francis
Publishers since 1798

UK	Taylor & Francis Ltd, 4 John St, London WC1N 2ET
USA	Taylor & Francis Inc., 1900 Frost Road, Suite 101, Bristol PA 19007

Copyright © Taylor & Francis Ltd 1995

All rights reserved. No part of this publication may be reproduced, stored in a retrieval system, or transmitted, in any form or by any means, electronic, electrostatic, magnetic tape, mechanical, photocopying, recording or otherwise, without the prior permission of the copyright owner.

British Library Cataloguing in Publication Data

A catalogue record for this book is available from the British Library
ISBN 0 7484 02624 (cased)

Library of Congress Cataloging in Publication Data are available

Cover design by Amanda Barragry

*Typeset by Mathematical Compositions Setters Ltd,
Salisbury, Wiltshire, SP3 4UF*

Printed in Great Britain by Burgess Science Press, Basingstoke, on paper which has a specified pH value on final paper manufacture of not less than 7.5 and is therefore 'acid free'.

Contents

Preface vii

Contributors ix

Section one: Fatigue in the transport industry

1. Trends and themes in driver behaviour research: the Fremantle Conference on 'Acquisition of skill and impairment in drivers' 3
   Laurence R. Hartley and Pauline K. Arnold

2. The road transport industry: a management perspective 15
   Owen Jones, Keith Edwards and Greg Weller

3. Managing driver fatigue in the long-distance road transport industry: interim report of a national research programme 25
   Anne-Marie Feyer and Ann M. Williamson

4. The driver fatigue and alertness study: a plan for research 33
   Deborah M. Freund, C. Dennis Wylie and Clyde Woodle

5. The role of fatigue research in setting driving hours regulations 41
   Narelle Haworth

Section two: The epidemiology of fatigue-related crashes

6. Road traffic crashes by region in Western Australia 51
   G. Anthony Ryan

7. Fatal accidents in Western Australia 59
   Peter J. Moses

8. Drugs, driving and enforcement 67
   Gavin Maisey

9. Tracing the problem of driver fatigue 73
   Peter Ingwersen

10. The interaction between driver impairment and road design in the causation of road crashes – three case studies 87
    Patrick J. Kenny

Section three: Countermeasures to the adverse effects of fatigue on driving

11. The road to fatigue: circumstances leading to fatigue accidents 97
    Dallas Fell

12. Motor vehicle deceleration indicators 107
    Pasha Parpia

13. An alcohol and other drug education programme for bus operators 121
    Tanya Barrett

14. The effect of profile line-marking on in-vehicle noise levels 127
    Peter Cairney

15. Detecting fatigued drivers with vehicle simulators 133
    Anthony C. Stein

Section four: Empirical analyses of the impact of fatigue

16. The road transport industry: the company driver's perspective 151
    Graham Derham and Colin Harwood

17. Methodological issues in driver fatigue research 155
    Ivan D. Brown

18. Alcohol, fatigue and human performance deficits: insights from additive factor logic 167
    Colin Ryan, Janet Greeley and Katherine Russo

19. Driver impairment monitoring by physiological measures 181
    Karel Brookhuis

20. Fatigue and alcohol: interactive effects on human performance in driving-related tasks 189
    David J. Mascord, Jeannie Walls and Graham A. Starmer

21. Developing a high fidelity ground vehicle simulator for human factors studies 207
    James W. Stoner

22. The development of a driver alertness monitoring system 219
    John H. Richardson

Section five: Theoretical considerations in research into driving

23. Dysfunctional driving behaviours: a cognitive approach to road safety research 233
    John M. Reid

24. Driving simulation in Australia: opportunities for research and application 249
    Thomas J. Triggs, Alan E. Drummond and John Stanway

25. Pro-active processing: the hallmark of expertise 259
    Kim Kirsner

Index 275

Preface

Fatigue is such a common experience that many people, who are able to, have organized their lives so as to minimise its more obvious adverse effects. The variety of techniques they use to achieve this are seldom reported in the scientific literature and therefore fail to contribute to the development of theory. For the vast majority of employees, however, fatigue prevention or limitation is not so simple. They have job demands to meet and fixed working hours to fill; therefore avoidance of the more serious consequences of fatigue becomes the joint responsibility of employer and employee. Unfortunately, it has been all too clear to both parties that countermeasures against fatigue are likely to involve unwelcome changes in job demands and/or constraints on hours of work, leading inevitably to conflict of interests and unreliable reporting of the causes and accidental consequences of fatigue. It is therefore unsurprising that the history of fatigue research is marked by sporadic, mainly government-funded attempts to solve very specific wartime problems and by less than successful working hours legislation imposed on particular industries by civil authorities ever mindful of the political risks of constraining commercial competitiveness.

These are not ideal conditions for researchers to develop theories of fatigue causation and prevention, nor to learn from the application of their theories to practical problems. Any progress made in this field has tended to result from the efforts of a few individuals, from a small number of technologically developed countries, who have persisted in studying the more practical consequences of fatigue among certain occupational groups, or from those more numerous psychologists and physiologists who have researched specific issues associated with fatigue – arousal, attention, vigilance, signal detection, sleep, circadian rhythms – largely for theoretical reasons, without using the term 'fatigue' at all and perhaps without even recognizing that their efforts were related to the phenomenon, as defined by those in the field.

Two events in particular can be perceived as having shaped current research interests in fatigue *per se* and identified it as a multidisciplinary phenomenon. First, Bartley and Chute's distinction between 'fatigue' and 'impairment', led to the definition of 'fatigue' as a psychological state in which individuals declare they are unable to continue performing the task in hand (either at all, or to their own satisfaction). 'Impairment' was thus left to be defined in terms of the various adverse performance and physiological changes that accompany fatigue, but which may not be precisely correlated with it. Second, there has been an increasing growth of interest over the past 30 years or so in the source and nature of circadian rhythms of physiological activation and their relationship with performance variation over time, to the point where chronobiology has now developed as an independent science. These two events have been important not only in relation to general theoretical interests in fatigue, but also more specifically in directing attention away from individual differences in fatigue susceptibility and towards the multitude of task and environmental contributions to fatigue causation. This had been particularly true among tasks such as driving and process control, where it has often been difficult, both experimentally and in accident analyses, to unconfound the interacting effects of time-on-task, time-of-day and sleep-loss. A problem for the advancement of knowledge in this field, however, has traditionally been the difficulty of integrating effort by the different multidisciplinary groups researching fatigue-related issues.

The organizers of the Conference in Fremantle, Western Australia, in September 1993, the proceedings of which comprise this book are therefore to be congratulated in bringing together a group of researchers covering the area of fatigue, as defined earlier, together with representatives of the transport industries and of the transport authorities having responsibility for the limitation of fatigue-related accidents. Seldom, if ever, has there been a comparable forum for the exchange of views on driving fatigue among individuals who could provide information on their problems, those who have the expertise required to research these problems, and those who need reliable evidence on which to eliminate or alleviate those problems.

It is particularly apposite, although unsurprising, that the conference which resulted in this book should have been held in Western Australia, where the need for shift-working by many drivers to cover the long distances involved in haulage operations, together with high temperatures and the open repetitive nature of the terrain, will inevitably exacerbate the causation of driver fatigue. What may be somewhat more surprising to potential readers, especially those with little previous experience of fatigue research, is the inclusion of 'driving simulation' as a central topic in the Fremantle conference. The main reason for this is the extreme difficulty of researching driver fatigue on the road. Apart from the obvious ethical objections to fatiguing drivers for experimental purposes, and therefore subjecting them to accident risk, it is virtually impossible to design an on-road study that provides adequate experimental control of traffic demand, time-of-day and time-on-task effects. Driving simulation offers a solution to such difficulties, while raising its own problems associated with the appropriate levels of 'realism' of display and control, which clearly need to be discussed in a multidisciplinary context, such as that provided by the Fremantle conference. A second reason for the inclusion of driving simulation in this context is the practical need to train drivers in the recognition of fatigue symptoms and in the employment of effective remedial measures. Third, there is a perceived need by the police for a valid method of establishing objectively that an accident- or incident-involved driver is actually impaired by fatigue. Finally, the need to evaluate the potential fatigue-provoking characteristics of new road designs *before* they are built is now recognized; clearly this can only be met by sophisticated simulation. All of these reasons are addressed to differing extents in this book.

There are chapters of interest here for the traffic authorities, the haulage operator and, especially, for the researcher. But this book does not represent the last word on the subject of driver fatigue: hopefully, it will serve to direct the attention of researchers and funding agencies alike to the vast amount of important work that remains to be done and the most effective ways of doing it.

Ivan Brown

Contributors

Arnold, Pauline, K.	Institute for Research into Safety and Transport, Murdoch University, Western Australia 6150, Australia
Barrett, Tanya	School of Occupational Therapy, Curtin University, Shenton Park, Perth, Western Australia
Brookhuis, Karel	Traffic Research Centre, University of Groningen, PO Box 69, 9750 AB Haren, The Netherlands
Brown, Ivan D.	Medical Research Council, Applied Psychology Unit, 15 Chaucer Road, Cambridge CB2 2EF, England – now retired: Ivan Brown Associates, 21 Swaynes Lane, Comberton, Cambridge CB3 7EF, England
Cairney, Peter	Principal Research Scientist, Australian Road Research Board Ltd, PO Box 156, Nunawading, Victoria 3131, Australia
Derham, Graham	Driver, Key Transport, Great Eastern Highway, South Guildford, Perth, Western Australia
Drummond, Alan E.	Monash University Accident Research Centre, Monash University, Clayton, Victoria, Australia
Edwards, Keith	General Manager, Key Transport, Great Eastern Highway, South Guildford, Perth, Western Australia
Fell, Dallas	Road Safety Bureau, Roads and Traffic Authority of NSW, PO Box 110, Rosebery, New South Wales 2018, Australia
Feyer, Anne-Marie	National Institute of Occupational Health and Safety (Worksafe Australia), PO Box 58, Sydney, New South Wales 2001, Australia
Freund, Deborah M.	Project Manager, Office of Motor Carrier Standards, Federal Highway Administration, USDOT, HCS-10, 400 7th Street, SW, Washington DC 20590, USA
Greeley, Janet	Department of Psychology and Sociology, James Cook University of North Queensland, Townsville, Queensland 4811, Australia
Hartley, Laurence R.	Institute for Research into Safety and Transport, Division of Psychology, Murdoch University, Western Australia 6150, Australia
Harwood, Colin	Driver, Gascoyne Trading, Jackson St., Bassendean, Perth, Western Australia

Haworth, Narelle	Monash University Accident Research Centre, Monash University, Clayton, Victoria, Australia
Ingwersen, Peter	Research Officer, Main Roads Department, Waterloo Crescent, East Perth, Western Australia
Jones, Owen	General Manager, Gascoyne Trading, Jackson St, Bassendean, Perth, Western Australia
Kenny, Patrick J.	University of Technology, Sydney, PO Box 123, Broadway, New South Wales 2007, Australia
Kirsner, Kim	Department of Psychology, University of Western Australia, Nedlands, Western Australia 6009, Australia
Maisey, Gavin	Consultant, WA Police Department, 2 Adelaide Terrace, East Perth, Western Australia 6004, Australia
Mascord, David J.	Department of Pharmacology, University of Sydney, New South Wales 2006, Australia
Moses, Peter J.	Main Roads Department, Waterloo Crescent, East Perth, Western Australia
Parpia, Pasha	Director, Cognitive Science Programme, University of Western Australia, Nedlands, Western Australia 6009, Australia
Richardson, John H.	HUSAT Research Institute, Elms Grove, Loughborough, Leicestershire LE11 1RG, UK
Reid, John M.	Director, WorkForce Solutions Pty Ltd, 7 Hammerdale Ave., East St Kilda, Victoria, Australia
Russo, Katherine	Department of Psychology and Sociology, James Cook University of North Queensland, Townsville, Queensland 4811, Australia
Ryan, Colin	Department of Psychology and Sociology, James Cook University of North Queensland, Townsville, Queensland 4811, Australia
Ryan, G. Anthony	Director, Road Accident Prevention Research Unit, Department of Public Health, The University of Western Australia, Nedlands, Western Australia 6009, Australia
Stanway, John	Transport Accident Commission, 222 Exhibition St, Melbourne, Victoria, Australia
Starmer, Graham A.	Department of Pharmacology, University of Sydney, New South Wales 2006, Australia
Stein, Anthony C.	Safety Research Associates, Inc., 4739 La Cañada Blvd, La Cañada, CA 91011, USA
Stoner, James W.	College of Engineering, Center for Computer Aided Design, 208 ERF, University of Iowa, Iowa City, Iowa, USA

Triggs, Thomas J.	Monash University Accident Research Centre, Monash University, Clayton, Victoria, Australia
Walls, Jeannie	Department of Pharmacology, University of Sydney, New South Wales 2006, Australia
Weller, Greg	Partner, D. & G.B. Weller, Perth, Western Australia
Williamson, Ann M.	National Institute of Occupational Health and Safety (Worksafe Australia), PO Box 58, Sydney, New South Wales 2001, Australia
Woodle, Clyde	Project Coordinator, Trucking Research Institute, 2200 Mill Road, Alexandria, VA 22314–4677, USA
Wylie, C. Dennis	Principal Investigator, Essex Corporation, Systems Effectiveness Division, 5775 Dawson Avenue, Goleta, CA 93117, USA

Section One:
Fatigue in the transport industry

1 Trends and themes in driver behaviour research: the Fremantle Conference on 'Acquisition of skill and impairment in drivers'

Laurence R. Hartley and Pauline K. Arnold

> It's hard work is mine; for I never have any rest but a few minutes, except every other Sunday, and then only two hours ... If I was to ask leave to go to church, I know what the answer would be – 'You can go to church as often as you like, and we can get a man who doesn't want to go there' ... If I'm blocked [in a traffic jam] I must make up for the block by galloping; but if I'm seen to gallop and anybody tells our people, I'm called over the coals ... It's not easy to drive a bus; but I can drive and must drive to an inch; yes, sir, to half an inch.
>
> Hibbert (1987: 655), quoting the recollections of an omnibus driver in the nineteenth century.

The costs and consequences of work, or other prolonged, concentrated activity have not always been of interest. These recollections of a transport driver in the nineteenth century, on the familiar problems of 'fatigue and speeding', probably reflect the experience of most transport workers in earlier times. In the twentieth century, attention has shifted to an emphasis on the demands for both mental and physical resources required of employees by their work. Working conditions have been improved by the reduction of working hours and the imposition of safety standards for all workers. Though influenced by these general improvements, the transport industry has, to the present, been little subject to industry specific restraint as regard to satisfactory working regulations. This state of affairs has arisen from the necessity for practical, regulatory latitude in the industry and from the difficulty of defining fatigue and measuring its consequences. Regulations to limit work in the transport industry were promulgated several years ago but at a time of inadequate knowledge of the several factors which impact upon accident risk. These regulations were devised to limit the exposure of employees to work but not to accidents. Since then it has become clearer to researchers that further issues need to be accommodated to limit the exposure of transport workers to accidents. In Chapter 17, Brown suggests that the main factors affecting transport drivers' exposure to the risk of fatigue-related crashes are not only the demands of the industry such as the requirement to engage in loading the truck before departure, adhere to delivery schedules, work unusual hours as regards the 'normal' diurnal cycle and regularity of driving schedule, but also individual differences between drivers, such as driving experience, personality, fitness, age and adaptability.

These themes are central to this book and it is hoped, they will form the focus for several future biennial conferences addressing the topic of fatigue in drivers, so appropriate in as large a continent as Australia, and its far flung transport industry. This book reports the proceedings of a conference held in Fremantle, Western Australia on September 16–17, 1993. The conference organizer was fortunate to receive a grant from the Road Trauma Trust Fund administered by the Traffic Board of Western Australia on behalf of the Minister of Police who also opened the conference. As a result of the grant, several distinguished international researchers were able to attend the conference and to present papers:

- Dr Ivan Brown OBE, former Assistant Director of the Medical Research Council's Applied Psychology Unit in Cambridge, UK and now of Ivan Brown Associates, presented a paper on methodological issues in studying fatigue. Ivan Brown has spent many years engaged in research into driver behaviour making very important contributions on many topics in the area including fatigue.

- Dr Karel Brookhuis, of the Traffic Research Centre at the University of Groningen in The Netherlands, presented a paper summarizing some of his research into correlates of driver impairment, including psychophysiological measures such as cardiac sinus arrhythmia.

- Dr Tony Stein formerly of Systems Technology Inc. (STI) and now Safety Research Associates, presented a paper outlining the use of STI's driving simulator configured to emulate a truck, to identify fatigue in drivers in Arizona.

- Dr Jim Stoner, Associate Professor of Civil Engineering, University of Iowa and Director of the Iowa Driving Simulator presented a paper overviewing the development and capabilities of this large-scale driving simulator, and reviewing some of its current uses.

Among other international speakers, Deborah Freund PE, a senior transportation specialist in the Federal Highway Administration's Office of Motor Carrier Standards of the US Department of Transport, presented a report of the planning and methodology involved in a large-scale study of driver fatigue being conducted by Dr Denis Wylie of the Essex Corporation and Dr Clyde Woodle of the Trucking Research Institute. Dr John Richardson of the HUSAT Research Institute in Loughborough, UK provided an overview of their research into driver alertness monitoring, a part of the European PROMETHEUS programme into developing new technologies to improve traffic flow and safety. Other participants included representatives of the transport industry (WA Road Transport Association), government research organizations, regulatory authorities and academic research workers.

This book is not intended as a comprehensive presentation of all current research programmes into fatigue. Rather, it aims to capture some of the key issues in current applied research programmes that are relevant to fatigue, including fatigue in the transport industry, the epidemiology of fatigue-related crashes, countermeasures to the adverse effects of fatigue on driving, empirical analyses of the impact of fatigue and some theoretical considerations in research into driving and fatigue. Its aim throughout is to maintain a link between researchers, road users and regulatory authorities, and to reflect the demands on the transport industry and its long-distance drivers such as those operating in Australia. To this end, Chapter 2 is contributed by managers in the road transport industry and Chapter 16 by transport industry drivers.

1.1 Section one: Fatigue in the transport industry

In Chapter 2, Jones, Edwards and Weller, describe the operation of the road transport industry in Australia, and Western Australia in particular. These authors are former drivers, and now managers, of road transport companies. Their contribution provides a backdrop to understanding the demands of the industry. It is complemented by Chapter 16 by Derham and Harwood, who are transport company drivers.

Several points in Chapter 2 are noteworthy. In Australia, as in many parts of the world, road transport is paramount, as compared to rail transport and coastal shipping. Accordingly, most manufactured and processed freight is transported by road from the urban to the rural parts of Australia and trucks return carrying produce and livestock. The return journey can exceed 7000 km. Much freight is also trucked east to west and back by truck in Australia. For economic reasons, prime movers tow 'road trains' of two trailers on busy roads, but three trailers can be towed in more remote areas such as the north-west of Western Australia.

Drivers may be either employed by a transport company, truck owner-drivers working under contract to a transport company, or independent truck owner-drivers finding freight to carry from many sources. The owner-drivers work under greater pressure than company drivers because of mortgage and service commitments to their vehicles, and they usually drive more frequently. Two crewing systems predominate on most freight routes. Over short to intermediate distances, 'solo' drivers predominate, sleeping in their trucks at night, if necessary. Over longer distances a crew of two – 'the two-up system' – alternate driving with rest and sleep about every four hours in the moving vehicle. Occasionally, a third crewing system, the 'staged system' is used, where drivers work for about eight hours in one vehicle and take their rest in a hotel, to return to driving another vehicle later; this system is unusual in freight transport but common in the coach transport industry. There is considerable discussion as to the 'better' system. Freight drivers usually prefer the 'two-up' system, especially over longer routes, because of the assistance a co-driver can provide and its potential to provide more flexible working and sleeping periods as compared to the solo system.

However, there are other considerations in choosing optimal truck crewing systems. Although two-up crews have more flexibility, solo drivers may obtain longer periods of uninterrupted sleep at more 'normal' times of night, not be required to drive at the 'lowest' point of the diurnal cycle, and will be undisturbed by movement of the truck. Accordingly, the impact of the diurnal cycle and of sleep quality upon the development of fatigue needs to be considered. It is possible that the 'staged system' is to be preferred over longer routes although economic considerations make it infrequent in the competitive freight industry.

In the freight industry some, but not all, truck drivers collect and deliver freight as part of a regular delivery service to rural customers. Meeting agreed delivery times forms a major constraint on all drivers' behaviour. As Owens, Edwards and Weller describe in Chapter 2, drivers employed or contracted by companies work to a regular schedule designed to allow them some discretion in their allocation of time to driving and rest and adequate time to complete the journey provided there are no mishaps and breakdowns. However, as the present author experienced, it is not impossible for a road train to have a flat battery a long way from help – then what can you do? Good planning of delivery schedules, permitting flexible management by drivers, is used as an argument by the industry for rejecting the need for further regulation of driving hours. This argument has some force as long as drivers can identify their state of fatigue and have the opportunity to take countermeasures. Conversely, the imposition of stricter driving hours regulations must not be used as a reason to modify driving schedules so that they do not permit drivers some discretion to rest should they feel fatigued.

When time is lost through breakdowns or other mishaps, serious time stress is placed on the driver to meet delivery schedules. Furthermore, independent owner-drivers, as these authors note, may take time to find a load to carry and then be required to deliver it to a more demanding schedule than company drivers. The independent operator is less likely to have adequate discretionary time for rest and sleep, and may be required to drive at the 'low' point of the diurnal cycle, just before sunrise and adhere to 'tighter' delivery schedules than company drivers.

Solving the problem of providing adequate discretionary time for sleep among drivers is complicated by the issue of whether drivers can always appreciate the point at which they are liable to fall asleep and need to pull off the road to recover. This is an important issue, for should some drivers be unable to recognize the onset of sleep, they cannot be thought to be driving carelessly if they fall asleep. In Chapter 16, Derham and Harwood, transport drivers, emphasize driving experience leads to an appreciation of the signs of impending fatigue and the development of compensatory strategies to deal with it. As diurnal rhythm research indicates, adaptation to regular routes and schedules permits better sleep on the road and recovery at home. These truck drivers confirm this and note that changing to a new driving schedule is accompanied by a phase of adjustment to its requirements.

Chapter 2, written by the transport managers, highlights the variety of freight transported, freight management practices and the versatility required of drivers. Moving animal livestock requires different practices to ordinary freight and may require drivers to navigate difficult terrain, such as floods and unsealed roads. Livestock transport may require adherence to a strict schedule of permitted stops for rest which is mainly dictated by the needs of the livestock rather than the driver. The environmental conditions under which Australian drivers operate may also be quite severe, especially when working in the north. Daytime temperature can exceed 45°C and night time temperatures can fall steeply. Humidity can be very high. These environmental conditions can make sleep difficult. Furthermore, the driving environment in many parts of Australia is extremely monotonous with some of the longest stretches of straight road in the world. These conditions probably impose particular demands upon the fatigued driver to stay awake.

Some of the challenges posed by these operational requirements are investigated in Chapter 3 by Dr Anne-Marie Feyer, from Worksafe Australia, who is engaged in a major study of factors contributing to fatigue, such as the demands and constraints under which drivers operate in the transport industry. In Chapter 4, Deborah Freund outlines the research approach to these issues adopted jointly in the USA by the trucking industry and the US Department of Transport. This research programme follows on from the classic work of Mackie and Miller (1978).

In Chapter 5, Narelle Haworth reviews evidence to support the impact of diurnal factors on fatigue-related accidents and cautions against overestimating the benefits of rest breaks which might usually be thought to be the most appropriate countermeasures to fatigue. She suggests that after several hours of driving, rest has only a temporary benefit rather than permitting complete restitution. The present Australian driving hours regulations specify maximum hours driven or worked. Thus, they take little account of the time of day of driving, activities during rest breaks or prior activity; all of which are highly likely to affect the experience of fatigue. Haworth suggests that any new draft regulations under discussion may need to consider not only these factors but also where the driving is carried out – for rural driving may be more fatiguing but the environment more forgiving than urban settings. She also points out that although car-driving hours are not regulated, not only do car drivers have fatigue-related crashes but also fatigue in the car driver may contribute to crashes involving trucks. In Chapter 16, Derham and Harwood confirm that the practices of car drivers constitute a major problem for truck drivers.

As an alternative to basing driving hours regulations on crash risk, Haworth (Chapter 5) suggests that performance based tests which have high face validity, such as discussed by Anthony Stein in Chapter 15, are worth considering. In-vehicle tests of the quality of driving are also discussed by Richardson in Chapter 22, and have much to recommend them since they may inform drivers of their deterioration of skills arising from any source including fatigue, which may be compounded for example, by sickness or alcohol consumption.

1.2   Section two: The epidemiology of fatigue-related crashes

The second section of the book deals broadly with the epidemiology of road crashes. The chapters focus primarily upon rural roads in Western Australia since they carry much freight, often have a gravel surface or shoulder and are characteristically monotonous for long distances possibly disposing drivers to reduced attentiveness and drowsiness. Anthony Ryan (Chapter 6) contrasts the distribution of crashes in urban, rural and remote areas of Western Australia. Ingwersen (Chapter 9) provides a finer grained analysis of single-vehicle accidents on rural roads in Western Australia and argues that they are under-reported. The majority of such accidents arise from leaving the road and striking the gravel verge or unforgiving objects, suggesting failures of attention among drivers – such as may arise from fatigue or drowsiness – are the cause. This suggestion is further supported by the success of painted edge lines in reducing single-vehicle accidents in rural settings. Moses (Chapter 7), Ingwersen (Chapter 9) and Ryan (Chapter 6) note that monotonous roads, even when well surfaced, appear to be associated with most fatigue-related crashes, and that these accidents are concentrated at weekends, at night, and among young drivers.

As expected about half of these rural fatalities are drawn from the 17–30 age group, about half occur at night despite the low traffic density, and by contrast to urban accidents nearly three quarters of rural accidents involve a single vehicle. These observations suggest that journeys at the start or end of vacations, made by in-experienced drivers unfamiliar with the rural environment, may be most likely to result in fatigue-related crashes. These points are reiterated by Dallas Fell in Chapter 11. They suggest that many rural road fatalities in Western Australia may arise from the unfamiliarity of vacationing urban drivers with rural road conditions including monotony, gravel surfaces, shoulders or verges. However, Moses' more detailed analyses (Chapter 7) show that nearly three quarters of rural fatalities are rural dwellers.

Given both these findings and the comments of Haworth (Chapter 5) on driving hours regulations for urban and remote areas, it is hard to decide whether urban or remote areas should require the more stringent driving hours regulations.

In Chapter 10, Patrick Kenny draws attention to the interaction between road design and fatigue and the contribution of activities prior to accidents attributed to fatigue. He reviews three case studies in which fatigue, arising from hours of driving, work prior to driving or time of day, may have contributed to accidents which took place in relatively unforgiving driving environments. Driving environments are designed with reference to criteria for average drivers. Kenny's results raise the question of whether extreme variations in the state of the driver, such as may result from fatigue, shift work, etc., need to be better catered for by road design criteria.

1.3 Section three: Countermeasures to the adverse effects of fatigue

This section discusses some countermeasures to the onset and impact of fatigue. In Chapter 11, Dallas Fell confirms that accidents involving fatigue are relatively more common in rural areas, more often result in serious injury than non-fatigue accidents and mainly involve solo, young, male, car drivers at night and at the weekend. Of surveyed drivers, 26 per cent reported accidents or near accidents attributable to fatigue.

Compared to other trips, journeys involving fatigue-related accidents were:

- more frequently for holidays and personal or family visits;
- less likely to start from home;
- more likely to commence late at night from work or from social and recreational places;
- more likely to occur within two hours of the start of the journey; and
- more likely to follow a night of inadequate sleep.

These latter findings confirm the impact of activities prior to the journey, and the time at which the journey is made, on the development of fatigue. Driving fatigue accrues not only from driving itself, but also from previous activities, poor sleep, and long hours of work prior to departure. Fell also found that drivers involved in fatigue-related accidents took fewer precautions against the onset of fatigue, such as resting and sharing driving, although most were aware of the potential benefits of these activities. This suggests the importance of increasing public awareness of the need to guard against fatigue.

If as suggested, the development of fatigue is accompanied by withdrawal of attention from the task of driving and impaired reaction time, some beneficial improvements could be made to vehicle signalling. Rear-end collisions are frequent, especially in conditions of reduced visibility such as fog and at night. In Chapter 12, Pasha Parpia describes a simple decelerometer signalling system which indicates to following vehicles, and to the vehicle's driver, the rate of deceleration due to braking. This device may be of assistance in reducing rear-end collisions due to drivers' reduced attention to the task of driving when they are fatigued.

The development of fatigue interacts with other factors which influence the general state of the driver. In a study of prolonged driving on a simulator, Nelson (1989) found that not only were responses to critical events adversely affected by alcohol, but also subjective complaints and aversion to the task increased over time, presumably reflecting the development of fatigue, but much more so under the alcohol conditions. Alcohol appears to hasten the development of the subjective symptoms of fatigue and may impair drivers' judgment about their state of fatigue. Accordingly, some responsible employers of professional drivers are embarking on educational programmes advising their drivers of the effects of alcohol, prior to introducing policies on alcohol and drug use by drivers. In Chapter 13, Tanya Barrett reviews the findings from one such alcohol and drug education programme provided for city bus drivers. The findings reveal some ignorance about the time course of the effects of alcohol and its after-effects. Given the findings from Fell's study (Chapter 11), that some drivers failed to adopt countermeasures to fatigue, better targeted programmes may be required to educate drivers about fatigue and countermeasures. Targeted educational programmes, which inform drivers of the symptoms of impending fatigue, might improve drivers' ability to recognize its signs and when to pull off the road, should their driving schedule permit. Such signs might be increasing micro-sleeps while driving.

If an increase in micro-sleeps precedes falling asleep at the wheel then it might be expected that they would be accompanied by increased lane drifting, of which the driver might often be unaware. In Chapter 14, Peter Cairney reports the results from a study which aims to alert drivers that they have drifted off the roadway, by introducing micro-sleep countermeasures such as 'rumble strips' at the edge of the road. Although the findings of his study show the benefit of rumble strips to car drivers they indicate that there is still some way to go in designing edge-lining which will alert truck drivers of their lane drift.

A further countermeasure to fatigue-related accidents is to identify fatigued drivers before an accident. This approach has been adopted by the Arizona Department of Public Safety; it is to develop a simple test of driver's performance which correlates with performance in a vehicle. In Chapter 15, Anthony Stein reports his findings of the relationship between the performance test, the Truck Operator Performance System (TOPS) and driving performance in a vehicle simulator. TOPS was successful in identifying about 50 per cent of drivers fatigued by extended driving. As would be expected from past research (Mackie and Miller, 1978; Hamelin, 1987), running off the road accidents, and standard deviations of lane position, heading error and vehicle speed increased significantly by the thirteenth hour of driving. These results again, point to the withdrawal of attention from the driving task as fatigue develops, perhaps accompanying micro-sleeps. The results are important, not only in confirming the findings from studies of real driving performance (Mackie and Miller, 1978) in the controlled environment of a driving simulator, but also in pointing to the possibility of devising a simple screening test (such as TOPS) for fatigued drivers, which can be administered at the roadside.

1.4 Section four: Empirical analyses of the impact of fatigue

This section presents some empirical research into fatigue, following an introduction to the job of transport driver by two company drivers in Chapter 16; Graham Derham and Colin Harwood emphasize several factors many of which, except the social skills required of drivers, are dealt with in the research papers. First, the drivers describe their job as requiring versatility. Transport drivers require a variety of mechanical, electrical and social skills to deal with co-drivers and customers, in addition to experience in driving large articulated vehicles. As they remark: 'It is not only about driving, being a driver.'

The manual work involved in loading the trailers before departure is recognized by drivers as contributing to fatigue on the road; accordingly the normal two-up driving shift of four hours may be initially shortened to take account of this. They recognize that fatigue-related accidents can happen at the start of a journey after hours of exhausting manual handling. Conversely, the modern truck requires little manual effort to drive and provides a satisfactory environment for sleeping though the impact of the diurnal cycle on sleep quality and latency remains a problem.

Derham and Harwood also emphasize the social aspects of driving. Good social skills are essential in dealing with fatigued co-drivers and the difficult customers to whom freight must be delivered after the drivers' long journey. Two-up drivers carefully select their co-drivers since they have to be compatible and have confidence in each other.

Many drivers prefer to drive solo for these and financial reasons. These company drivers emphasize their job is less stressful than that of being an owner-driver because their duties are more regular and they do not have the financial and maintenance commitment to their vehicle that owner-drivers have to consider.

Becoming a driver entails an element of selection; drivers who cannot adapt to working schedules, are uncongenial or unreliable drivers, or with poor accident records are not likely to remain in the industry. It is tempting to suggest that these attributes reflect personality characteristics, such as neuroticism, known to be important in job satisfaction and performance. The authors assert that experience at the job of long-distance driver enables the development of adaptive and compensatory strategies to identify, minimize and deal with fatigue. Even the ability to obtain a satisfactory sleep in the moving vehicle is an acquired ability, they claim. Experience enables drivers to self-monitor their state of fatigue, pace their activities and sufficient flexibility in the delivery schedule, barring breakdowns and other mishaps, allows drivers to take preventive action at its onset such as swapping drivers or pulling over to rest. Accordingly, Derham and Harwood, not unreasonably, suggest that accidents attributed to fatigue arise from a failure to develop adequate strategies to deal with it.

Derham and Harwood suggest that regularity of driving schedule is very helpful in permitting drivers to adapt to the rigours of the job and to make the best use of their time off for recovery and domestic commitments. Although contract owner-drivers often drive to the same schedule every week, company drivers work two 'long' weeks, one 'short' and have one week off in each month. This contrast raises the question of how long a break to recover should drivers have between trips? Too long a break may interfere with their adjustment to the driving schedule; too short a break may not permit adequate recovery from the trip. The authors note that changes of driving schedule, such as a new route may require, can take time to adjust to and are initially fatiguing. Also, age takes its toll of the ability to cope with the stress of working at the low point of the diurnal cycle, just before sunrise; accordingly older drivers avoid those hours of driving.

In Chapter 17, Ivan Brown tackles the methodological issues in studying fatigue. He notes that despite a wealth of research, driver fatigue remains a problem partly due to its 'invisibility' in accident causation and partly because of resistance to the imposition of further constraints on the transport industry. He attempts a broad definition of fatigue which encompasses the subjective costs of sustaining performance under fatigue and incorporates the notion of a 'disinclination to continue the task because of perceived reductions in the efficiency with which it is conducted'. This definition contains the crux of how drivers, such as Derham and Harwood, handle fatigue; good self-pacing of activities and self-monitoring enables drivers to minimize fatigue and discontinue driving before their task efficiency is over-impaired. However, Brown notes that it is not clear yet whether all drivers are unequivocally aware of when they are about to fall asleep at the wheel, nor which internal cues the competent driver uses to determine the state of fatigue. He also notes that feedback to drivers about their performance by the development of 'advanced transport telematics' should overcome this problem, if suitable measures can be found.

In his review of the causes of fatigue, Brown draws out many of the issues which other authors, such as the drivers Derham and Harwood, have identified as contributory factors. For example, the contribution of time-at-the-wheel to the development of fatigue is agreed to be probably small in comparison to the contribution of time-of-day or inadequate sleep. Drivers agree with Brown that factors which mediate the development of fatigue include age, regularity of working schedule, length of experience at the job, an adaptable personality and physical fitness. Brown notes that despite past research, it is unclear whether some of the consequences of fatigue reflect skill impairment, behaviour compensating for fatigue or a change in the subjective risk attached to driving events. As noted in a study of transport driving to the north-west of Western Australia, (Hartley *et al.*, 1994) lateral lane deviations increased on monotonous roads at night when there was little other traffic; however, drivers steered quite adequately when driving into town a little later, suggesting that drivers can vary their steering criteria according to circumstances in order to minimize mental work load. Nevertheless, in the trial reported previously by Stein (Chapter 15), lateral lane deviation increased over driving time despite controlled traffic density on the driving simulator, suggesting that fatigue results in increasingly poor control skills rather than increased use of compensatory strategies. In conclusion, Brown suggests there has been a failure to address adequately driving hours regulations in the past, because of the methodological difficulties attached to studying fatigue, and that many of the issues discussed above remain to be answered satisfactorily for regulatory authorities.

In Chapter 18, Colin Ryan provides an empirical analysis of the effects of alcohol on memory search tasks. He argues that fatigue research should follow a similar model of analysing operation-predicated behaviour rather than task-specific behaviour such as truck driving. In Chapter 20 David Mascord and colleagues draw attention to many of the similarities of the effects of alcohol and fatigue and their potential to interact synergistically, in causing impairment in, for example, lateral lane position error. These authors confirm the point made earlier, that older drivers are more susceptible to the effects of fatigue even over a three-hour period of testing, and their decline in performance is potentiated by alcohol.

Several chapters have foreshadowed the introduction of intelligent vehicle highway systems such as the European Communities 'advanced transport telematics'. They have, for example, discussed some of the requirements for monitoring the state of the driver and providing appropriate feedback to drivers. In the final three chapters in this section, development work towards introducing and evaluating intelligent vehicle highway systems is described. In Chapter 19, Karel Brookhuis reviews the relationship between physiological and performance measures of drivers and concludes that a combination of such measures could be used to develop a driver impairment monitoring device to combat the adverse effects of fatigue by providing feedback to drivers of their state of driving capability.

Driving simulators form not only a very useful diagnostic tool for fatigue as Anthony Stein has described (Chapter 15), but can also be extremely useful for researching the problem, since they permit experimental manipulations to be carried out which would be impracticable on the road. They also have the added advantage of providing detailed measures of driver performance which can only be accomplished by elaborate instrumentation of road vehicles. In Chapter 21, Jim Stoner provides a detailed picture of the development requirements for a high-fidelity driving simulator at the University of Iowa. This simulator provides unparalleled fidelity for the driver and is currently being used to undertake virtual prototyping of vehicles, infrastructure and control systems, and to study driver performance and fitness to drive. This facility has the potential to be of great use to understand fatigue, evaluate road designs for inducing fatigue and tolerance of the variability in drivers' states.

In Chapter 22, John Richardson from the HUSAT Research Institute describes a collaborative programme of research with the Ford Motor Company to develop new technologies for better vehicle management such as collision avoidance, vision enhancement and diagnosis of fatigue. Richardson discusses the requirement for such intelligent systems to adapt to the state of the driver's alertness. In the research programme, not only numerous vehicle parameters, but also indicators of the state of the driver under various conditions of fatigue induced by sleep loss, are used to train a neural net to recognize instances of fatigue in the driver. Richardson makes the point that his data show fatigue to be a phasic state; his sleep-deprived subjects alternated between periods of alertness and drowsiness with increasing frequency during prolonged driving. These data also bear on the question raised by Brown (Chapter 17), as to whether drivers always know they are about to fall asleep at the wheel. In several instances, Richardson's subjects continued driving with eyes closed and no effective control of the vehicle. The final stage of the research will be to optimize the delivery of feedback to drivers of their state of alertness.

1.5 Section five: Theoretical consideration in driving research

The final section discusses cognitive theories of driving. A better understanding of fatigue and all aspects of driving, will be facilitated by the development of sound theories of driving performance, which incorporate not only the skills required but also the motivational factors affecting drivers. In Chapter 23, John Reid presents a model of driver behaviour which includes the development of appropriate schemata for the driving environment and the driver's affective state. By and large most driving research has ignored the impact of the driver's affective state on driving performance; however, some recent research has emphasized its importance (Gulian et al., 1989; Hartley and El Hassani, 1994). In Chapter 24, Tom Triggs also discusses the paucity of adequate theorizing, particularly in the area of young driver behaviour, and suggests this has led to the failure to deal with the problem of the over representation of young drivers in crashes. In particular, the absence of an adequate theoretical understanding of driver performance has meant that road safety interventions are poorly targeted at specific problems. This state of affairs can now be more directly addressed by using driving simulators to improve understanding of the process of driving, such as the types of information required for safe driving. Detailed studies of drivers on simulators shows young drivers' performance has features in common with fatigued drivers; their performance is more variable than that of experienced drivers. Simulators provide many opportunities to improve the training of young drivers by providing them with a controlled and structured driving environment, immediate feedback on their performance and potentially, unlimited opportunities to develop expertise in hazard perception.

Finally, in Chapter 25, Kim Kirsner compares several models of skill acquisition in driving. He proposes that structural models do not provide a productive platform for research in this area, and that two characteristics, involving dynamic changes in representations, and pro-active processes, respectively, provide a more appropriate platform for theoretical development.

References

Gulian, E., Matthews, G., Glendon, A.I., Davies, D.R. and Debney, L.M., 1989, Dimensions of Driver Stress, *Ergonomics*, **32**, 585–602.
Hamelin, P., 1987, Lorry drivers' time habits in work and their involvement in traffic accidents, *Ergonomics*, **30**, 1323–33.

Hartley, L.R. and El Hassani, J., 1994, Stress, violations and accidents, *Applied Ergonomics*, **25**, 221–30.
Hartley, L.R., Arnold, P.K., Smythe, G., Kirsner, K. and Hansen, J., 1994, Indicators of fatigue in truck drivers, *Applied Ergonomics*, **25**, 143–56.
Hibbert, C., 1987, *The English: A Social History 1066–1945*, London: Guild Publishing.
Mackie, R.R. and Miller, J., 1978, Effects of hours of service, regularity of schedules and cargo loading on truck and bus driver fatigue, Report 1765-F Human Factors Research Incorporated, Santa Barbara Research Park, Goleta, California.
Nelson, T.M., 1989, Subjective factors related to fatigue, *Alcohol, Drugs and Driving*, **5**, 193–214.

2 The road transport industry: a management perspective

Owen Jones, Keith Edwards and Greg Weller

2.1 General freight transport by company drivers: Owen Jones

The majority of freight moved over long distances in the state of Western Australia is on scheduled services; that is, a truck can pull one or two trailers in the busier areas of the state and three trailers in the more remote areas (a road train) and leave Perth (the state's capital city) on a certain day each week and arrive at its destination on a regular day each week. These services run in accordance with their set schedule irrespective of whether the truck is fully loaded or leaves Perth largely empty. A major transporter of freight in Western Australia is Gascoyne Trading, a division of Wesfarmers. Gascoyne Trading commenced its operations in 1924 to carry produce from Carnarvon, 900 km north of Perth, southwards to Geraldton to link up with the rail service then operating between Geraldton and Perth (Figure 2.1). Today, that produce is delivered direct from Carnarvon to the markets of Perth. Back in the 1920s, the trip took two days; today it takes 11–12 hours.

Road trains made up of two or three trailers are the most common method of long-distance road transport, pulling combinations of freezer, pantechnicon or flat-top trailers. Freezer trailers can have three or more types of freight in the one unit, that is, freezer goods, chiller goods and dry goods. A two-trailer road train measures up to 36 m in length with a gross mass of 79 tonnes. A three-trailer configuration has an overall length of 53 m and a gross mass of 115.5 tonnes. The prime mover used to haul these combinations is usually a six-wheel 400–500 hp engine long-bonnet truck.

It should be noted that the transport regulations currently in force in Western Australia allow a prime mover to pull two trailers only between Perth and Carnarvon and Perth and Meekatharra, 660 km/Wubin 260 km north of Perth. North of these centres, three trailers are allowed. Operationally this means that three prime movers each pulling two trailers, go as far as Carnarvon (or Wubin Meekatharra). Two prime movers each pulling three trailers continue north leaving one prime mover to return south to Perth with another load. As mentioned before, Gascoyne transports just about anything, travelling north. When it comes to the trip back south, to Perth, a large part of what is transported is 'back loaded' freshly grown produce from Carnarvon, Kununurra and, to a lesser extent, the Broome area. In addition, there is meat and hides, scrap metal, fish produce, especially frozen prawns for export from Carnarvon, Exmouth and Darwin.

Gascoyne Traders has greatly increased its scheduled services since those early days and now operates between Perth and all major towns in the north-west of the state, moving 2000 tonnes of freight a week. Most freight is brought into the company's Perth depot in the suburb of Bassendean and delivered to the client's premises – a depot-to-door service. Currently Gascoyne has 26 drivers on scheduled services, 12 company vehicles and 12 subcontractors. The present fleet of trailers numbers over 180.

Towns in the north-west area are serviced twice a week, sometimes as often as three times each week. In this way, clients know exactly when the Gascoyne vehicle will be

Figure 2.1 A schematic map of Western Australia, distances in kilometres and locations described in the text.

in town with their supplies, whether those supplies be a box of matches, a carton of beer, a car tyre, a piece of machinery or a household appliance. A typical scheduled service for Gascoyne is the service it operates from Perth to the holiday resort of Broome. Given the distance from Perth, 2232 km, the Broome run is a two-up operation; that is, two drivers rotate between driving and sleeping at about four-hour intervals to drive the truck to Broome. The drivers arrive at the Perth depot at 10 a.m. of the day of departure. As company drivers, their first task is to supervise the loading of the (two) trailers so they know exactly where the freight of each client is located. Contract drivers would themselves load one of their two trailers. Driver's departure time is between 5 p.m. and 7 p.m. the same evening, following completion of their paper work. Arrival time in

Broome is scheduled for the second morning following the evening departure. That is, if the vehicle left Perth on Tuesday evening, it would arrive on Thursday morning in Broome and then the freight would be delivered to customers by the driver. During the 36 hours of the Perth–Broome scheduled service, the drivers would average around 75 kph.

The company trucks do not exceed the speed limit of 100 kph for road trains. In fact, they are road geared so they are unable to exceed 100 kph. The hills of the trip are encountered between Perth and Northampton. It takes between five and six hours to arrive at Geraldton, 425 km north of Perth, and another hour-and-a-half to reach Northampton.

Drivers are encouraged to stop every 100 km to carry out a tyre and load check. There is some variation between drivers concerning how often they stop. They all have a particular hill or preferred road stop. If one of the two-up drivers starts to become tired, he will swap with the other driver who takes over the driving of the vehicle.

It should be noted that there are actually three company drivers in rotation to service each two-up driving schedule, one at home and two away with the vehicle, working two weeks on (paid) and one week off (not paid). However, contract drivers who drive the same route may work every week. With a two-up operation, the truck only stops while the drivers carry out their safety checks or stop for a meal. There is no need for the vehicle to stop while a driver sleeps. Whoever is not driving is able to sleep in the sleeper cab, a separate area located behind the driving position. On his week off, a driver usually prefers to take the time off. If he wants extra pay, he can come to work and do a short run, such as taking trailers to Carnarvon ready for the three-trailer operation north as shown in Figure 2.2. Occasionally, very occasionally, he will work around the vehicle loading dock. Drivers generally do not like working on the dock.

Figure 2.2 A three-trailer road train in the Pilbarra.

The precise driving–sleeping rotation is worked out between the particular pair of drivers. There are 'owls' and 'fowls' – some drivers prefer to drive during the day; others prefer night driving. A driver may drive for a couple of hours, then swap. He may, on the other hand, drive for six hours or so before handing over to his mate. In other words, the drivers self-regulate in accordance with their personal characteristics and preferences and in relation to the particular route they are driving.

Long-distance drivers working for Gascoyne and other established, reputable companies are all very experienced. Drivers currently employed by Gascoyne average 15 years experience. This average figure includes the drivers who are currently being trained, which means the more experienced drivers have been driving long-distance vehicles for between 25 and 30 years. Younger drivers are brought up through the ranks, starting with learning to load, driving vehicles in and out of the yard, then they will be taken on a shorter run, say to Carnarvon, as an offsider, gradually building their experience.

With 'single' or 'solo' driving, as opposed to two-up driving, the longest run is Perth to Port Hedland, a distance of about 1750 km. The time allowed for this journey is 36 hours, with the driver leaving Perth on Tuesday evening and arriving in Hedland on Thursday morning around 6 a.m. Again, the driver self-regulates his driving pattern, probably doing slightly longer driving spells on the way up, when he is fresh. The actual driving time is about 22 hours so there is no pressure on the driver. He has ample time to sleep during the 36-hour trip.

Drivers on the scheduled services described should not experience fatigue. Then again, what is fatigue? There is no particular definition. Everybody, when they work hard grows tired. The drivers know the job well; they are experienced and can pace themselves. When a driver becomes tired he stops driving, or swaps with his co-driver if driving two-up. There is a clear difference between professional driving and commuter or leisure driving. This is not always understood by the general public when there are debates about fatigue and driving hours. A professional driver is clearly superior to the amateur driver, driving to and from work or taking the family out for a picnic. A professional becomes attuned to regular driving, knows their limitations, adjusts appropriately to different weather conditions and road environments and drives accordingly.

As far as Western Australia is concerned, and especially when you are undertaking scheduled services, self-regulation, rather than prescribed driving hours, is the way to go. The drivers know when they are tired and need to stop driving. That is part of being a professional. They are not under any pressure to drive past their physical limits. It is drivers who are under pressure to keep driving when they should stop for a break or a sleep, that have the problems.

Gascoyne's accident rate is almost nil especially with company drivers as opposed to subcontractors. With respect to the reservations expressed about two-up driving in parts of eastern Australia, it should be borne in mind that the industry there has not had a great deal of experience with two-up driving. Drivers in Western Australia have been brought up with two-up driving whereas in the eastern states the drivers are often resistant to co-driving because their experience has largely centred on single driving. Two-up trips in the eastern states of Australia are often overnight between capital cities and the pressure is on the drivers to meet very tight delivery schedules. In Western Australia, the drivers support two-up driving – there are two drivers to load and unload and deal with any problems that arise and it adds to discipline. A driver is less likely to take risks when he has a co-driver on board.

On the issue of alcohol consumption, most Gascoyne drivers are social drinkers. They do not have an alcohol problem. The drivers value their job, they are highly paid, averaging $AUS70–90 000 per year. Gascoyne does not tolerate any misuse of alcohol by their drivers; the drivers know they face instant dismissal. Neither is there any particular problem with 'pill-popping' among Gascoyne drivers. Drivers are not placed under the sort of pressure which produces that problem. The clients of the company are not demanding unrealistic delivery times. The company sets the transit and delivery timetable. Gascoyne informs its clients when they will need to have the freight ready for loading at the dispatch depot in order to reach a destination at the time the client requires.

In contrast, the situation which is often the case in the eastern states, is that clients are placing more and more pressures on transport companies who in turn pressurize their drivers. The motivation for clients to apply pressure is to reduce the amount of stock they hold in store.

Turning to the type of prime movers used in Western Australia's long-distance sector, Foden and Mercedes have largely given way to USA brands, i.e. a Mack or a Kenworth. Gascoyne is progressively standardizing its fleet with Kenworths; other companies prefer Macks. Subcontractors engaged by Gascoyne, who own their own prime mover and pull the company's trailers, appear to have a preference for Volvos. Gascoyne uses a mix of company drivers and subcontractors, presently 50/50 in number. The mix is determined by the level of capital investment and a satisfactory return on funds invested being achieved. The advantage of the subcontractor from the company's perspective, is that it is the subcontractor who has the capital invested in the prime mover and, less often, in the complete truck and trailer combination. On the other hand, there is the disadvantage with subcontractors in that a subcontractor can be under greater pressure in the event of his prime mover breaking down and the expense of repairing it, plus all the worry of losing income if his vehicle is off the road for a period of time. In the event of a subcontractor breaking down, however, Gascoyne will do everything to assist him and to ensure that the company's clients are not disadvantaged. A company driver, of course, does not have these worries. The equipment he is driving is owned by the prime contractor, i.e. his company, and his wage packet is regular and waiting for him.

2.2 Specialized freight transport: Keith Edwards

Aside from the transport of general freight (as outlined above by Owen Jones), movement of specialized items for the mining, earthmoving and manufacturing industries is provided by various heavy haulage operators, such as Key Transport. There is demand for the transport of equipment such as locomotives, front-end loaders, bulldozers, excavators, dump trucks or electrical equipment as shown in Figure 2.3. Unlike general scheduled freight services, specialized transport can vary. This requires the operator to react quickly to customer needs. Drivers therefore are required at times to 'pack their toothbrush' at short notice for trips which can last up to 12 days. These urgent jobs can result in pressures arising from operations staff endeavouring to obtain the necessary permits and vehicle escorts from the regulatory authorities.

Items to be transported can weigh 20–160 tonnes and are placed on low loaders ranging from bogie and tri-axle configurations through to 128-wheel platform trailers. Prime movers used are specifically rated for these types of moves and may in fact, require a second prime mover to be used either in tandem or push–pull configuration. Trip times can vary according to load type, weight, dimension, distance and weather conditions.

Figure 2.3   A specialized transport operation.

Generally, all specialized movements of this nature require co-operation with various regulatory authorities to ensure, among other things, courtesy and safety to other road users. Permits are required from the Main Roads Department of Western Australia and it has conditions on it concerning supervision for the load over certain bridges, whether or not special approval is needed from the State Electricity Commission or whether police escorts are also required.

An example of where a police escort is required all the way would be moving a locomotive from Perth to Port Hedland, 1600 km north of Perth, on a platform trailer 32 m long, with a height of 5.5 m and a width of 3.5 m. To obtain the services of a police escort a booking must be made well in advance. Movement takes place during daylight hours only (although special permission under certain conditions can be obtained for night travel). Communication by drivers to all escorts is essential to ensure a smooth and safe transit.

Loads may also be carried in convoy, requiring further communication between drivers. Movements of this nature become very much team oriented to ensure safety and courtesy to other road users. All movements are carried out by one driver per truck. Drivers are provided with sleeping accommodation in their trucks which are air-conditioned. In the Perth metropolitan area, drivers must be off the road by 7.30 a.m. and cannot resume until 9 a.m. In the afternoon, drivers must be off the road by 4.30 p.m. and cannot resume until 6 p.m. provided it is still daylight. These restrictions are to avoid peak traffic times.

In a 12-hour working day, drivers would not exceed 10 hours of driving. I cannot see that the drivers engaged on heavy haulage should suffer any fatigue problem. Following each job, the vehicles are serviced and any other work such as tyre changes carried out, ready for the next job. In the event there is a quiet period, the prime mover is used on some other company work in order to ensure maximum utilization of the company's equipment.

2.3 Livestock transport: Greg Weller

This section attempts to explain some of the complexities of livestock transport. In particular, the transport of cattle from the Kimberley region in the north of Western Australia, as this requires virtually non-stop driving for 2400 km to the metro area. The vehicle configuration typically would be a three-trailer six-deck unit capable of moving around 120–150 fully grown animals with an all-up weight of 60–70 tonnes, giving a gross combination of 115–120 tonnes. Because of the requirement that all cattle coming from the Kimberley are free of parasites, they are unloaded and dipped in insecticide at Broome. They are held there for 24 hours and redipped. This gives drivers an opportunity to service vehicles and check everything out. They are then in a position to have a good 12-hour rest before reloading. Once the cattle have been loaded (which usually takes about 40 minutes) little time is lost in starting out.

In this type of transport, it is the driver's responsibility to consider his welfare and that of the stock and to strike a happy medium. In recent years, the Great Northern Highway has seen a good spread of road houses built which enables a driver to check his stock and obtain refreshments at the same time. For the first 100 km, generally two checks are made of the cattle. This allows the driver to check on how the animals are travelling and decide if he can extend the stop to 3–4 hours. Also, of course, the tyres are checked at each stop and visual appraisal of equipment is carried out. These stops generally take 10 minutes. Meal breaks are usually taken night and morning with light snacks in between.

The fact that showers are available at roadhouses has made long-distance transport of cattle a little easier. A shower and a meal with around an hour break, helps the driver combat fatigue and the cattle benefit as well from non-movement of the trailer. Other factors that assist drivers is the provision of a refrigerator in the cab for drinks and snacks, air-conditioning of cab and sleeper with separate controls and a sleeper bunk that allows the driver – even big drivers – to stretch out.

The provision of two drivers is another factor that is often used when a lot of stock have to be moved. This system can be used to educate novice drivers by allowing them limited hours of driving and gaining an understanding of the requirement of cattle transport. Because the driver has full control of his truck during the trip, he is able to make a decision on his need for rest. A short 1–2 hour sleep would be fairly normal; however, depending on the problems encountered, a longer rest may be necessary.

The key factor in long-distance cattle transport is moving the stock from point A to point B in as short a time as possible, without putting other road users at risk. A normal trip from Broome to Perth would take around 38 hours. This means the cattle are delivered to Perth in good condition with as little weight loss from dehydration as possible.

Concerning shorter hauls of stock, i.e. 400–1500 km, sheep are often the livestock being transported. These are moved in purpose-built four-deck crates weighing around 10.5–11.5 tonnes. Their capacity is around 400 fully grown sheep giving a gross combination of 42.5 tonnes or 76 tonnes in a two-trailer configuration.

Again the same cab comforts apply as in long-distance work; however, because of the geographical spread of sheep, there are different applications of techniques. The driver has a longer loading period with sheep and more physical exertion. Also, he may be required to travel on roads poorly served by roadhouses. This generally means that the medium- and short-haul livestock transporters require more rest to cope with fatigue.

The drivers considered to be the best or 'top' operators are those who drive between Broome and the more remote cattle stations in the north-west of Western Australia, often on the roughest of the roads. These drivers are prepared to take the hard work, the rough conditions, and the truck rollovers. They are paid for it, but it is tough going. One such operator spoke about a station 48 km off the main road which was a nine hour trip in and out.

> Driving in the remote, northern parts of Australia is a different world. It is too hot to sleep, so you drive. You sleep when you are buggered. I have been driving in the Northern Territory when it is 43°C, with a load of valuable stud cattle. You cannot keep stopping to rest, unloading and reloading the cattle. If you do that – make unscheduled stops – you have to get your permits altered. The permit nominates in advance where you are going and where you are stopping. These permits are issued by the Department of Agriculture specifying the cattle are disease free. You cannot risk picking up diseases along the route, remembering a deck of stud cattle can be worth $120 000, compared with ordinary store cattle which might be worth $9000 per trailer deck.

So, livestock transport is complicated; it is not like general freight. Everything has to fall into place. So the driving task has to be self-regulatory, otherwise it just would not work. When a driver has had enough, he will pull over. The records speak for themselves; there are very few accidents in livestock transport and the ones that do happen are not caused by a driver falling asleep. It is equipment malfunction and road conditions that have been the cause.

> In the north of Australia, you are usually driving on gravel or worse. You have to watch for thunderstorms. The road might be all dusty but there might have been a thunderstorm 50 or a 100 km away. There is nothing more startling at night, going along steady, steady, and suddenly seeing froth on the road. Actually, it is good if there is froth because you can see it. But when the water has stopped black across the road – I tell you what – it is hard to see. Suddenly you are in three feet of water or worse. And with sudden downpours I have seen water around Karratha, 1600 km north of Perth, rise faster than I can drive through it: four inches of rain in 20 minutes. When I started through it was six inches deep, by the time I was through it was four feet deep. The driver behind me was stuck in the middle. You have no way of knowing just how much rain has fallen.
> Another problem up north is that a lot of the roads you are driving on are not fenced. So, you have the extra hazard of cattle on the road. You know there is going to be cattle all over the place. They will just sit on the road. There is a spot just near Sandfly. It is unbelievable there. It is almost impossible on a bitumen road to see some colours of cattle, if they are on the road, unless they turn to face you and you can see their eyes. Another problem is when someone has hit a beast in front of you and you cannot see it. I have been over a dead beast and it gives you a hell of a shaking up. In a lot of cases a car can stop. It can pull up quick enough, but there is no way a road train can.
> I have been carting livestock since the late 1960s including horses and stud cattle going north and market cattle coming south – horses from Perth to Port Hedland, Broome, Derby, Fitzroy Crossing, Hall's Creek and cattle, just about everywhere, to whatever station has bought them. You turn up at eleven at night or one in the morning to make the delivery, off load and keep going. Over the years there have been some pretty funny places I have had to meet people. If they had only bought two bulls they would meet you somewhere on the road, at the start of the spinifex bush or somewhere. You would not go right in to deliver them. They would sometimes have to travel 100 km in the dark to meet up with you.
> With the sophisticated gear you have these days, it is not as though you are driving something that is physically wearing you out. If you are having a hard time, that keeps you awake. There is nothing better for staying awake than having a tyre blow out. You are mad with it and you can stay awake all night.
> Most trucks travel around 85 kph. I used to sit on 85 kph all the time. Trucks are often geared to a limit of 90 kph. Fully loaded with big bullocks, your cannot travel faster than this. The only things that pass you at night are the express trucks. The fact that once you are south of Carnarvon or Meekatharra you have to stop to change down from towing three

trailers to two is one reason we have a very good safety record. Once you stop, you can have a shower, and something to eat and rest. The biggest hazard I find is the bloody heat reflection signs on bridges. A blue background with a reflector is no problem – but those damn black and silver – well, at night the reflection back from your lights is absolutely disgusting. It blinds you, it is like being hit with a welding flash.

Any introduction of driving hour restrictions would result in extra cost on the industry which it could not afford. Margins already have gone down and down. And the first thing you have to look at is that there is only a five-month driving season, starting around May. You have to be pretty efficient to make a profit. It can look nice picking up $12000 for a trip but you have spent $6000 for fuel alone. And if you are unlucky with the wind, blowing on the angle against the trailer with a height of 4.6 m, it is like hitting a brick wall. Up goes your fuel consumption and down goes your margin.

I do not think I have seen anyone, the genuine operators, in livestock transport suffering a fatigue problem.

3 Managing driver fatigue in the long-distance road transport industry: interim report of a national study

Anne-Marie Feyer and Ann M. Williamson

3.1 Introduction

The job of long-distance driving is likely to [be one that] demands long periods of sustained attention [in a monotonous environment]. Considerable evidence exists documenting the [difficulties of such tasks present for] the human operator (Davies, Shackleton and Parasuraman, 1983; Folkard and Monk, 1985; Krueger, 1989).

To date, most of the solutions advanced to overcome driver fatigue have concentrated on placing maximum limits on the number of driving and/or working hours and minimum limits on the number of hours of rest in a given time period. The major advantage of such regulations is that they provide an enforceable way of limiting the number of hours worked, relative to the number of hours of rest obtained.

As an approach to managing driver fatigue, however, regulation of the driver's hours of driving, working and rest has a number of critical shortcomings. First, current regulations place limits on consecutive hours of work and rest irrespective of time of day. They do not take into account the contribution of human circadian rhythms to alertness nor of sleep physiology (Moore-Ede, Campbell and Baker, 1988; Hertz, 1988). Thus, current regulations do not deal with fatigue caused by irregularity in rest and duty hours.

Another problem with the approach is that current regulations are not derived from an empirical research basis. There is wide variation from country to country in the form and the time periods specified, without any scientific basis for the efficacy of the chosen regime. Yet, despite considerable variation in permissible hours of service requirements, driver fatigue remains a serious problem universally, and the impact of hours of service on fatigue remains largely unresolved (Hamelin, 1987; US Department of Transportation, Federal Highways Administration, 1990; MacDonald, 1984).

Finally, controlling the hours of working, driving and rest fails to take account of inter- and intra-driver variability. It is likely that different patterns of work will result in different levels of fatigue for the same driver, and that drivers will differ in the tolerance they develop to the inevitable fatigue that results from driving. Clearly, consensus about the real limits of duration of driving and duration of work is lacking. Currently there are no research-based alternatives to existing working hours regulations. What is becoming increasingly clear, however, is that fatigue needs to be viewed not only in terms of long hours of driving but rather as part of the whole pattern of work and rest. Certainly, the amount of work relative to the amount of rest is an important factor but other factors are likely to be of equal importance. The nature of the rest obtained, including its quality and timing needs to be considered, as does recovery time between

---

*The views expressed in this paper are those of the authors and do not necessarily reflect those of the National Occupational Health and Safety Commission.

trips. It also seems likely that non-driving activities such as loading will influence the level of fatigue experienced by drivers. These are all factors that have been shown to be relevant to fatigue in other industrial settings (Rosa *et al.*, 1990) and in truck driving (Mackie and Miller, 1978).

The aim of this project was to identify possible strategies to improve management of driver fatigue in the long-distance road transport industry. However, there has been an attempt to broaden the view of the possible range of solutions to the problem of fatigue to include a range of work practices. To achieve this aim, the study was designed to have two stages; the first stage was designed to gather information about a range of strategies, both current and potential, which might be helpful in reducing driver fatigue; and the second was designed to evaluate a small number of strategies judged to be most successful on the basis of the first stage of the study. The first stage of the study has been completed (Feyer *et al.*, 1993; Williamson *et al.*, 1992). This chapter, reporting the findings of the first stage of the study, examines the relationship between current operational practice and driver fatigue in the freight sector. The second stage of the project, currently in progress, involves objective on-road evaluation of some of these relationships. The basis of this work is also described.

3.2 Stage 1: Method, results and discussion

Method

A questionnaire was designed to obtain information about fatigue from drivers. The questionnaire included details of the driver's driving experience, details of current employment and working conditions, as well as details of their last trip and their last working week. Drivers were also asked to report about their experiences of fatigue including the effects of fatigue on their driving, factors that contribute to their fatigue and how they deal with the problem. They were also asked for their views about the helpfulness of a range of strategies that could be used to combat fatigue.

The questionnaire was designed to be self-administered or administered by interview. Self-administered questionnaires were distributed mainly through companies and at truck stops in all mainland states. Interviews were conducted at truck stops. Population-based stratification of the sample was not possible because no reliable statistics are available about the structure of the industry. Consequently the sampling strategy adopted was to survey the industry as widely as possible in order to include as many views as possible.

In all, 960 long-distance truck drivers participated in the study. Of these, 658 completed the questionnaire by self-report, and 302 subjects were interviewed.

For the purposes of analysis, the sample was classified in two ways. Drivers in the sample were classified according to the type of operation driven on the last trip: single, staged or two-up. In single operations, a single driver took the load from point of origin to point of destination. In two-up operations, a team of two drivers took the load from point of origin to point of destination. In staged operations, one driver took the load to a change-over point, where the load was then taken to the next or final point by another driver with the first driver returning to his point of origin. Among drivers of single and staged operations, some drivers reported about round-trips rather than their last one-way long-distance trip. Although the data for two-way and one-way trips are presented separately, analysis of these data has revealed that the subgroups within operations do not differ substantially.

Drivers in the sample were also classified on the basis of employment status into employee or owner-driver, and within these groups drivers were further classified according to their relationship with particular sized companies. Small companies were defined as those with fewer than 10 trucks, medium companies as those with 11–50 trucks and large companies as those with more than 50 trucks. Owner-drivers could be classified as independent owner-drivers if they did not work for any particular company.

Results and discussion

Overall, the results showed that the majority (77.5%) of truck drivers believed fatigue to be a problem for the industry but not a major personal problem (see Table 3.1). However, most (59.7%) reported feeling fatigued before the fourteenth hour of driving, and most commonly in the early hours of the morning. Significantly, most drivers reported that their driving was adversely affected, with remarkable consistency across drivers in the ways in which driving was reported as being affected. Fatigue appears to make drivers slower to react and poorer in steering and gear changing. Drivers were also consistent in the factors nominated to contribute to fatigue while driving. Dawn driving, poor roads, long driving hours, poor weather and having to load as well as drive were all reported as commonly contributing to fatigue.

Differences were evident, however, in fatigue experience among drivers in different sectors of the industry (see Table 3.2). Single drivers and two-up drivers reported fatigue as occurring more often and as a greater problem for them than did staged drivers. Of the three operation types, two-up drivers reported among the highest levels of fatigue. Yet, staged drivers became fatigued earlier in the trip than either two-up or single drivers.

Analysis of the operational conditions reported by drivers working under different driving operations provided some useful insights into why these differences in fatigue experience occurred. Single and two-up drivers reported doing longer trips than staged drivers, especially so in the case of two-up operation. Two-up and single drivers also reported working longer weekly hours than staged drivers. These differences in workload could be expected to result in greater fatigue among single and two-up drivers. Single and two-up drivers were also more likely to be involved in loading activities, which may also have contributed to their experience of greater fatigue.

These pressures on two-up and single drivers were offset to some extent, however, by other operational conditions reported by these drivers. Compared with staged

Table 3.1 Attitudes to and experience of fatigue reported by all drivers

	Rating % of all drivers (N = 969)
Extent of the problem	
At least a substantial personal problem	34.9
At least a substantial industry problem	77.5
Frequency of the problem	
At least occasionally	36.0
On at least half of trips	84.6
On the last trip	50.6
Onset of fatigue	
Before the 14th hour	59.7

Source: Williamson et al., 1992.

Table 3.2 Attitudes to and experience of fatigue reported by drivers of each type of operation

	Single (1-way) N = 696	Single (2-way) N = 123	Two-up N = 43	Staged (1-way) N = 33	Staged (2-way) N = 26
Extent of the problem					
At least a substantial personal problem	38.5	32.8	28.6	15.1	19.2
Frequency of the problem					
On at least half of trips	50.8	37.4	49.9	24.3	19.2
On the last trip	59.3	49.6	50.0	34.4	44.0
Onset of fatigue					
Mean hours after starting work	13.0	10.7	18.6	7.8	11.1
(s.d.)	(10.7)	(5.5)	(18.1)	(2.8)	(3.2)

Source: Williamson et al., 1992.

drivers, two-up and single drivers were more likely to schedule their own start and finish times, and less likely to have start times in the early hours of the morning. These characteristics combined to provide greater flexibility for single and two-up drivers to organize the structure of their trips and thereby stave off fatigue for longer periods than could staged drivers. Coupled with the inflexible structure of their trips, staged drivers also reported more night work in the past week than other groups. It is likely that chronic exposure to driving at the most vulnerable times of the day contributed to cumulative fatigue among staged drivers, manifest as succumbing to the fatiguing effects of driving earlier in their trips (Williamson et al., 1992). The higher levels of fatigue reported by two-up drivers is likely to reflect that the vastly greater distances these drivers cover ultimately outweighs the benefits of flexibility in trip scheduling.

Similar conclusions could be drawn from the analysis of the influence of operational conditions to be found under different employment arrangements (Table 3.3). Independent owner-drivers and employees of large companies reported similar levels of fatigue, with both groups being low reporters of fatigue compared with other employment categories. Although independent owner-drivers reported experiencing fatigue on most trips more often than employee drivers working for large companies, they also reported being able to go further in their trips before experiencing fatigue (see Table 3.3). Yet, independent owner-drivers reported doing significantly longer trips than any other employment category. Again, the reasons for this pattern of fatigue experience appeared to be due to other characteristics of independent owner-drivers operations. Specifically, independent owner-drivers reported greater flexibility in organizing the structure of their trips, making it easier for them to manage fatigue.

Taken together, these results provided some clear directions for strategies for evaluation in the second stage of the study. Overall, the data confirm the importance of timing of rest and work, not just the length of the work period. Drivers who can achieve some personal control over the structure of their trips, despite doing longer trips, were lower reporters of fatigue than those drivers who did not have such flexibility. It also appears that staged and two-up operations, two current operational practices specifically designed for better management of fatigue, may not be achieving their intended outcome. Both practices provide a relief driver to allow the truck to cover longer distances while attempting to provide the driver with effective rest. Yet two-up drivers were among the highest reporters of fatigue. Staged drivers, although not high reporters

Table 3.3 Attitudes to and experience of fatigue reported by drivers in each employment category

	EMPLOYEES N = 737			OWNER DRIVERS N = 223			
	Small N = 215	Medium N = 143	Large N = 379	Independent N = 56	Small N = 43	Medium N = 49	Large N = 72
Extent of the problem							
At least a substantial personal problem	36.8	46.9	32.5	26.0	46.6	40.9	35.2
Frequency of the problem							
On at least half of trips	48.4	56.5	38.2	56.9	67.4	48.9	42.3
On the last trip	57.0	56.2	53.7	54.9	66.7	66.0	56.3
Onset of fatigue							
Mean hours after starting work	14.7	12.6	10.4	16.2	15.3	14.6	11.9
(s.d.)	(13.4)	(10.7)	(6.0)	(14.6)	(8.1)	(15.7)	(5.7)

Source: Williamson et al., 1992.

of fatigue, reported fatigue early in their trips, suggesting that their work schedules are particularly taxing in the short term.

3.3 Stage 2: directions and progress to date

The second stage of the study has been designed to investigate the impact of flexibility for the driver to control trip organization, and two-up and staged driving, in objective on-road evaluations. The results of the second stage are expected to provide on-line evidence of driver functioning during different ways of operating.

Method

In general terms, the methodology involves volunteer company drivers and owner-drivers driving their regular truck under normal operational conditions. Three comparisons have been planned. First, staged driving as it is currently operated will be compared with single driving. Staged driving is intended to reduce fatigue by providing a relief driver at a changeover point, rather than have a single driver take the load to its destination. Therefore, the essential starting point for understanding the impact of staged operations on driver fatigue is to evaluate it against the usual alternative.

Similarly, two-up driving is intended to allow the vehicle to cover vastly longer distances by having a team of two drivers share the driving, rather than have a single driver take the load over the same distance. Again the essential starting point for understanding the impact of two-up operations on driver fatigue is to compare it to the usual alternative, single driving. In this way, the impact of work practices demanded by two-up operations as they are currently implemented can be directly evaluated.

The third comparison planned is specifically aimed at evaluating whether flexibility in current work/rest regulations would be of benefit to drivers in reducing fatigue. Driving a regular route according to the schedule demanded by the current regulations will be compared with driving the same route according to a preferred schedule.

The range of measures chosen for study reflect that both the causes of fatigue and its manifestation are likely to be multifaceted. Figure 3.1 shows the range of factors that are likely to affect driver fatigue. In designing the second stage of this study,

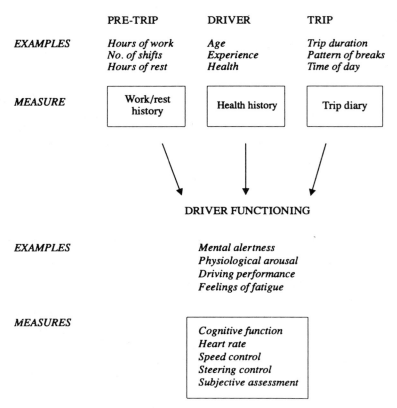

Figure 3.1 Possible sources of fatigue and the measures of these included in the study.

measures have been selected which allow the range of possible factors to be evaluated so that their relative contribution to the problem, and interactions between factors, can be determined. Specifically, as Figure 3.1 shows, both factors outside the particular trip as well as aspects of the trip are likely to have an impact on fatigue, and have therefore been included.

To determine the impact of recent work history, drivers will provide detailed diaries of their work and rest schedule in the week before each trip being evaluated. Each driver will also provide a detailed general health history. This history will obtain information about relevant medical conditions and also about lifestyle factors such as exercise, smoking and alcohol consumption. In this way, a detailed profile of the driver's status before the trip being evaluated will be obtained.

To evaluate the impact of characteristics of the trip, detailed diaries will be obtained for the trip being evaluated. As well as basic information about trip length and timing, details of work and rest on the trip will be obtained, the number and timing of breaks taken, activity during breaks and the like.

Evaluation of the impact of each type of trip on driver functioning under operational conditions will be monitored. Instrumentation to evaluate the impact of the driving regimes has been designed to obtain these data in real time without infringing on the driver's attention or comfort and maintaining the flexibility of using the driver's regular truck.

Driver fatigue, as Figure 3.1 shows, can be manifest in various outcomes. These outcomes are not necessarily contemporaneous and may fluctuate as a result of various factors, for example strategies by the driver or changes in demands of the driving task. In this study, all four dimensions will be evaluated in order to describe the impact of the driving regimes on the driver as fully as possible.

Firstly, cognitive functioning of the driver will be evaluated before, during and after the trip using part of a portable battery of tests developed by the authors. The tests from the battery to be included are the critical flicker fusion test, a simple manual reaction time task and a vigilance task performed over a 10-minute period. The battery also includes an unstable tracking task based on the Jex task (Smith and Jex, 1986). In addition, an auditory signal requiring a vocal response will periodically occur throughout the trip. This task will provide information about changes in cognitive function during each trip uncontaminated by having the driver stop and changing the environment, factors which could be expected to result in some re-arousal when the driver stops to perform the test battery tasks.

Physiological functioning will also be evaluated. Throughout the trip heart rate will be monitored to obtain inter-beat interval and inter-beat variability. Cardiac function is considered to be systematically related to autonomic arousal, and will therefore provide a continuous measure of driver functioning at this level.

Although physiological and cognitive functioning underlie all performance, including driving performance, evaluation of changes in driving provide the most direct measures of the impact of driver fatigue. Therefore, this dimension has also been included in the study. The variability in the control of steering and speed will be measured throughout each trip. Deterioration in both of these aspects of control of the driving task have been related to driver fatigue. These measures will allow direct evaluation of the impact of different driving operations on driving outcomes.

Finally, subjective assessments of fatigue levels will be obtained by self-report questionnaire before, during and after each trip. While operational phenomena like physiological arousal and cognitive functioning are amenable to objective assessment, fatigue remains, essentially, an experiential phenomenon, which can only be assessed by asking the individual. Furthermore, it is on the basis of their subjective appraisal that drivers are likely to make decisions about taking rest to alleviate fatigue.

Progress to date

The instrumentation for the second stage of the project has been finalized. A data logger has been built for collection and storage of continuous input from steering, speed and heart-rate sensors, as well as delivery of the episodic on-road reaction time signal and collection of the vocal response time data. The data logger also includes a real-time clock, allowing all data collected to be time stamped. In this way, changes in each of the dimensions can be related to each other.

Currently, the impact of staged driving and the impact of flexibility for the driver to control trip timetables are being evaluated. Data collection is presently under way in this stage of the project using an inter-capital city route on the east coast.

Conclusion

The aim of this project was to identify possible strategies to improve management of driver fatigue in the long-distance road transport industry in Australia. In the first stage, the industry was surveyed to determine the nature of the pressures on drivers from different sectors of the industry, their impact on fatigue and whether any specific methods for reducing fatigue were being used by any groups in the industry that could be applied to other settings within the industry. The results of the first stage highlighted a number of factors which deserved further investigation, either because they were promising as strategies to reduce fatigue or because they were already being used in the industry to reduce fatigue but did not appear entirely successful. The second stage of the study, currently underway, was designed to investigate these factors in on-road operational trials. Taken together, the results are expected to provide input to evaluation of current operational and regulatory practices in the industry to manage fatigue, and to allow recommendations for improved management of the problem.

Acknowledgments

This project was funded by the Australian Federal Office of Road Safety.

References

Davies, D.R., Shackleton, V.J. and Parasuraman, R., 1983, Monotony and boredom, in Hockey, R. (Ed.) *Stress and Fatigue in Human Performance*, Chichester: John Wiley and Sons.

Feyer, A-M., Williamson, A.M., Jenkin, R.A. and Higgins, T., 1993, Strategies to combat fatigue in the long distance road transport industry. The bus and coach perspective, Report No CR 122, Canberra: Federal Office of Road Safety.

Folkard, S. and Monk, T.H., 1985, *Hours of Work*, Chichester: John Wiley and Sons.

Hamelin, P., 1987, Lorry drivers' time habits in work and their involvement in traffic accidents, *Ergonomics*, **30**(9), 1323–33.

Hertz, R.P., 1988, Tractor-trailer driver fatality: the role of non-consecutive rest in a sleeper berth, *Accident Analysis and Prevention*, **20**(6), 431–9.

Krueger, G.P., 1989, Sustained work, fatigue, sleep loss and performance: a review of the issues, *Work and Stress*, **3**(2), 129–41.

MacDonald, N., 1984, *Fatigue, Safety and the Truck Driver*, London: Taylor & Francis.

Mackie, R.R. and Miller, J.C., 1978, Effects of hours of service schedules and cargo loading on truck and bus driver fatigue, Contract No DOT-HS-5-01142, Washington, DC: National Highway Traffic Safety Administration.

Moore-Ede, M., Cambpell, S. and Baker, T., 1988, Falling asleep behind the wheel: research priorities to improve driver alertness and highway safety, in *Proceedings of Federal Highway Administration Symposium on Truck and Bus driver Fatigue*, Washington, DC.

Rosa, R.R., Bonnet, M.H., Boozing, R.R., Eastman, C.I., Monk, T., Penn, P.E., Tepas, D.I. and Walsh, J.K., 1990, Intervention factors for promoting adjustment to nightwork and shiftwork in *Occupational medicine: state of the art reviews*, Philadelphia: Hanley and Belfus.

Smith, J.C. and Jex, H.R., 1986, *Operational manual critical task tester MK 10*, Hawthorne, California: Systems Technology, Inc.

US Department of Transportation: Federal Highways Administration, 1990, *Hours of service study: Report to congress*.

Williamson, A.M., Feyer, A-M., Coumarelos, C. and Jenkins, T., 1992, Strategies to combat fatigue in the road transport industry. Stage 1: the industry perspective, Report No CR 108, Canberra: Federal Office of Road Safety.

4 The driver fatigue and alertness study: a plan for research

Deborah M. Freund, C. Dennis Wylie and Clyde Woodle

4.1 Introduction

The Federal Highway Administration (FHWA) is developing a comprehensive research and technology transfer programme to address the issues of commercial motor vehicle (CMV) driver proficiency. This programme encompasses driver medical qualifications, training, and fitness for duty. The core of this plan will be a series of high quality, and scientifically sound, research projects. The results of these projects will form the basis for regulatory changes to the US Department of Transportation's Federal Motor Carrier Safety Regulations, which govern the interstate movement of people and property, that will be more technically defensible, better reflect the realities of the operating environments, and, to the extent possible, be based on performance standards.

The Driver Fatigue and Alertness Study is the first major field research project within the 'fitness for duty' element of the programme. It is also the first to collect data from CMV drivers behind-the-wheel since the mid-1970s.

4.2 Background

Under the Truck and Bus Regulatory Reform Act of 1988, the US Congress directed the Department of Transportation to conduct research to determine the relationship, if any, among federal hours-of-service regulations for operating CMVs, operator fatigue, and the frequency of serious accidents involving CMVs.

The FHWA, an agency within the Department of Transportation responsible for, among many other things, the safety of CMV transportation, held a symposium on 'Truck and Bus Driver Fatigue' in November 1988, to discuss what was known about fatigue and fatigue-related accidents and to propose future research. The conference brought together experts from the motor carrier industry, the scientific and medical communities, law enforcement, and public policy. The Driver Fatigue and Alertness Study was a direct result of the symposium recommendations.

4.3 Study goals and objectives

The goal of the study is to observe driver fatigue and loss of alertness, and developing countermeasures to combat it, within the framework of a realistic driving environment. The study has several objectives:

- to establish measurable relationships between driver activities and physiological, psychological, and subjective indicators of fatigue and loss of alertness;
- to identify and evaluate effectiveness of alertness-enhancing countermeasures that drivers may legally, safely, and practically use, and develop educational materials and presentations; and

- to provide a scientifically sound basis to determine the potential for revisiting the current Hours of Service requirements in the Federal Motor Carrier Safety Regulations.

The study is a major scientific research effort to gather information on a broad range of interrelated items in the driver–vehicle environment:

- driver performance and vehicle operating parameters;
- objective and subjective measures of driver psychological and physiological state; and
- vehicle operating environment (such as cab temperature and air quality).

It consists of a set of field experiments designed to replicate a range of carrier operations, performed under 'real-world' conditions, i.e. revenue-producing trips:

- **Baseline set**: these were daytime schedules with regular starting times and 10 hours of driving.

  Adding approximately two hours on-duty-not-driving time to the workday (for vehicle inspection, administrative tasks, etc.) and one hour of off-duty time (meals, breaks) gave the driver roughly 10 consecutive hours off-duty time prior to the next starting time, 24 hours later.

- **Rotating** or **operational schedules**: these started earlier each day to maximize distance travelled while adhering to the 10-hour driving limit and providing the minimum 8-hour off-duty required under US regulations.

  As in the baseline set, the on-duty-not-driving and off-duty periods were included in the daily schedule. Each successive starting time was therefore three hours earlier than the day before, introducing a circadian effect.

- **13-hour driving schedules**: while longer than the USDOT regulations currently permit, these schedules may promote increased driver alertness by keeping the driver's work and rest cycles closer to 24 hours.

  These runs were done in Canada under its provincial rules. As in the two other schedules, the on-duty-not-driving time and off-duty time were included in the drivers' workdays. This resulted in a 16-hour work shift. Providing the minimum eight consecutive hours of off-duty time prior to the next driving period required under Canadian regulations maintained a constant 24-hour cycle between starting times.

The second part is a study of the feasibility of countermeasures:

- to review devices that directly monitor the driver;
- to review devices that monitor vehicle operational parameters; and
- to identify things that drivers do to enhance their level of alertness, delay the onset of fatigue, or mitigate its effects.

The study is a venture developed with the help of, and which will continue to involve, the motor carrier industry. The FHWA has organized technical consultation sessions to solicit viewpoints on perceived relationships among the scientific goals of the study, the declared research needs of various concerned parties, the proposed experimental protocols, and the opportunities to apply those protocols toward collecting information during 'real world' motor carrier operations – revenue-producing runs. The meetings have brought together representatives from the trucking industry, drivers, law enforcement officials, scientists, and policy experts.

The study is also a public–private partnership, and an international one. The Trucking Research Institute of the American Trucking Associations' Foundation is funding a significant portion of the data collection and analysis effort. Transport Canada is cost-sharing in the extended-hours portion of the data collection. The American Trucking Associations, the National Private Truck Council, the International Brotherhood of Teamsters, and the Owner-Operator Independent Drivers Association have provided considerable input in the public forums. These organizations paved the way for recruiting motor carriers and drivers, and provided ongoing technical and operational support to the research effort.

4.4 What the study is not

On the other hand, there are several things that the study is not.

- It is not an exhaustive analysis of conditions affecting driver alertness over the entire motor carrier industry. We are limited by time, logistical, and funding constraints. We planned to collect the information during actual, revenue runs. As we solicited motor carrier firms and their drivers, we were mindful of the need to balance the need for 'representative' data with causing as little disruption as possible to the carriers.

- It is not a basis for rulemaking proceedings leading to wholesale changes in the Hours of Service regulations. We are developing a research baseline, but we cannot expect that we are replicating the full range of commercial motor carrier operations in a single study.

- It is not a basis for a set of regulations concerning the application of alertness-enhancing countermeasures. Our research plan in the countermeasures area includes a review of scientific literature and current practices in industries where shift work is performed, as well as a survey of some 500 drivers to find out how they keep alert.

- It is not an attempt to compare behaviour resulting from fatigue with behaviour resulting from other potential impairments (e.g. alcohol, drugs and disease).

4.5 Funding and administration

Research funding for this study was made available for the fiscal year following the 1988 symposium. A contract was awarded to the Essex Corporation, Columbia, Maryland, in September 1989. The work is being performed by Essex's Human Factors Research Laboratory, Goleta, California.

The initial 'kickoff' meeting was held in November 1989, to review the first draft of the formal study design. After initial revisions, the FHWA sponsored a technical consultation on May 31–June 1, 1990, inviting a smaller group of motor carrier industry and other experts. Numerous well-supported technical recommendations towards revising the study design and augmenting data collection equipment were made to the FHWA. Most of those recommendations were incorporated through a major contract modification, signed in April 1991.

In September 1990, a second, companion, contract for collection and analysis of brain wave and other physiological data from the same driver group was awarded to the Trucking Research Institute of the American Trucking Associations Foundation. The subcontracts for this extremely specialized and complex research were awarded in January 1992. Essex Corporation is responsible for organizing the physiological data collection and integrating the results with the driving performance data; Scripps Clinic and Research Foundation, La Jolla, California is responsible for reducing and interpreting the physiological data.

In the summer of 1993, an international research agreement between the US Department of Transportation (USDOT) and Transport Canada was expanded to include sharing of resources for the conduct of the portion of field data collection activity in Canada. In return, the USDOT will share data products and analyses with Transport Canada.

4.6 Technical aspects of field data collection

While it is difficult, if not impossible, to measure a driver's 'fatigue', it is certainly feasible to measure many of its mental, physical, and physiological elements. Its characteristics include:

- decreased alertness
- decreased vigilance/watchfulness
- increased information processing and decision-making time
- increased reaction time
- more variable and less effective control responses
- decreased motivation
- decreased psychophysiological arousal (measured by changes in body temperature, brain waves, heart action and nervous system activity).

Changes in these states can be measured from four standpoints:

- **Driving performance**

    Lane tracking, steering wheel movement, information processing speed, eye-hand co-ordination and critical tracking, reaction time

- **Driver internal state**

    Heart rate, heart rate variability, and vagal tone

    EEG and EOG

    Patterns of eyelid response

    Body temperature

- **Driver subjective state**

    Standardized fatigue rating scale, rest/sleep logs

- **Sleep quality**

    EEG, EOG, blood oxygen saturation level

Most of these items can be recorded continuously while the driver is behind the wheel, or sleeping. Data on the driver's ability to process information, as well as self-assessments of subjective state, were recorded periodically while the vehicle was stopped.

Five instrumentation suites were built and used. Each consisted of a lane-tracking device, forward-facing and driver-facing video cameras, a video recorder with screen splitter, steering wheel angle and speed sensors, temperature and humidity sensors, and a portable vagal tone monitor (Figure 4.1). A portable computer, keyboard and display were used to record instrumentation outputs and to administer the periodic information processing and performance tests. Each driver wore scalp and facial electrodes for EEG and EOG recording and three chest electrodes for ECG recording. Each driver used an infra-red tympanic thermometer at intervals for temperature recording. EEG and EOG signals were recorded on an ambulatory recorder which also provided signal conditioning. A portable gasoline generator powered all electrical equipment.

It is important that data from different sources be correlated so that a comprehensive assessment may be made of changes in patterns of driver alertness. Internal clocks on all data collection instrumentation were synchronized. All data were time stamped at the time of recording, and all clock readings were concurrently recorded on the same media to permit precise evaluation of any differences among clocks.

The study sought driver volunteers with at least two years of experience driving the type of CMV they would operate in the study. The drivers had to be 25–65 years of age and not have histories of sleep disorders nor of alcohol or drug abuse. The drivers had to be willing to drive the specified routes and schedules and to sleep in the sleep-lab quarters provided. They also had to permit the ECG, EEG and EOG hookups and to submit to scheduled and random tests for alcohol and drug use.

Drivers at most of the candidate US motor carriers ('less-than-truckload' motor carriers that utilize nationwide networks of terminals to consolidate freight) have labour

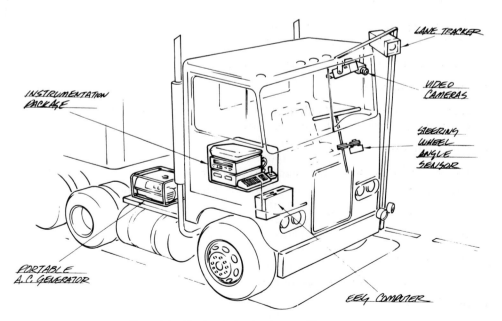

Figure 4.1   Truck cab showing layout of instruments.

representation through the International Brotherhood of Teamsters (IBT). It was therefore extremely important to earn the support of the national organization, as well as that of the locals where participating drivers were based. The IBT had a representative on the Technical Consultation Group who participated actively and recommended several adjustments to the study plan. The study team maintained close contact with the IBT throughout the data collection preparation period.

The IBT agreed to support the general goals of the research programme, and to work with their locals, if necessary, to address administrative concerns. It became apparent that there was a conflict between seniority-based assignments of schedules and routes and the study's need for a range of ages of participating drivers. The participation of the local's officers was central to an agreement being reached between the motor carriers' management and the drivers' union representatives to adjust driving assignments for the study.

Participating drivers received an orientation led by the principal investigator and various members of the study team, and signed an informed-consent form. They took a physical exam, and were trained to perform the computer-based tests. Drivers were compensated at their regular hourly wage for time spent preparing for the study, taking secondary performance tests, and 'laying over' at the sleep-lab.

4.7 Countermeasures

'Countermeasures' is a broad term relating to monitoring of driver behaviour, or to monitoring of vehicle operational parameters related to execution of driving-oriented tasks, as well as actions that drivers can take to enhance their level of alertness, delay the onset of fatigue, or mitigate its effects.

Countermeasures to loss of alertness may be classified as 'alarms', which activate after a driver's alertness level has diminished, and as 'maintainers', that are designed to keep alertness levels from dropping below safe levels (Mackie and Wylie, 1991). Alarms are further categorized by types that monitor driving performance and types that monitor physiological state. Maintainers can be controlled by the driver, or controlled by an external entity, such as motor carrier management, or by a government agency's regulations.

Driver-controlled alertness maintainers have received very little, if any, scientific scrutiny. This class of countermeasures include auditory stimulation, both passive (listening to a radio or cassette tape) and active (conversing on a citizens'-band radio), mental games, control of cab heat and humidity, use of stimulants (caffeine, nicotine, others), autoregulation (biofeedback), and brief stops for meals, vehicle checks, or naps. They will be the main focus of the study's countermeasures survey.

Externally controlled alertness maintainers comprise two categories: institutionally based and environmental. The former include government regulation of permissible driving and on-duty hours relative to driving, minimization of irregular day/night driving patterns, rest stops required by regulation or by motor carrier operational practice, and driver and company training. Environmental alertness maintainers – or alarms, depending on how the driver encounters them – include highway-infrastructure items such as raised pavement markers, rumble strips and flashing warning lights. As noted earlier, institutionally based countermeasures, in the form of revised hours-of-service regulations (if warranted), are a desired outcome of a comprehensive and technically sound research programme, of which this study is but one component. We are monitoring research by others on environmental alertness maintainers.

4.8 Status

While the revisions to the study plan were being formalized during the spring and summer of 1991, motor carriers were recruited for participation in the field data collection exercise. In July 1992, a pilot test of equipment, software, and data collection procedures was conducted. That fall and winter, adjustments were made to the hardware and procedures, drivers were recruited from the participating companies, and five more sets of test equipment were built.

The over-the-road field data collection activity was completed in December 1993. Three major for-hire motor carriers – Yellow Freight Systems and CF Motor Freight in the US, and SLH Transport, a wholly-owned subsidiary of Sears Canada – had a total of 80 drivers 'on the road' under the baseline and operational/rotating variations of the 10-hour schedules (in the US) and the 13-hour daytime and nighttime schedules (in Canada). The US runs took place from mid-June to early August, and the Canadian drivers' activity began in late September and ran through the first week in December.

As of early 1995, data analysis is underway. The massive database covers over 200 000 miles of driving. It includes some 5000 hours of video data, 10 000 hours of physiological recordings (heart rate and brain wave), and 700 megabytes of real-time truck computer records. Data quality is very high, especially considering the challenging conditions of a truck travelling 500 miles (more in Canada) per day. The research team estimates that the size of the summary database will be 1600–2400 megabytes. All analysis is being performed on 80386- and 80486-class microcomputers. Data are being handled within a relational database management software package, selected to facilitate utilization of a powerful microcomputer-based statistical analysis package.

During mid-1995, the countermeasures survey of up to 500 drivers will be performed. The objective is to learn about, and assess, safe, legal and effective actions that drivers can take to enhance their level of alertness, delay the onset of fatigue, or mitigate its effects.

4.9 Closing

The FHWA, Transport Canada, and research contractors, the Essex Corporation of Goleta, California, and the Trucking Research Institute of the American Trucking Associations Foundation, along with representatives from the motor carrier industry and scientific and technical organizations, are working together in a research partnership to achieve the goal of observing driver fatigue and loss of alertness, and developing countermeasures to combat it, in a realistic driving environment. We believe that the knowledge we gain through this research will help make roads safer for all the drivers who share them.

References

Mackie, R.R. and Wylie, C.D., Countermeasures to loss of alertness in motor vehicle drivers, *Proceedings, 1991 Human Factors Society Annual Conference*.

5 The role of fatigue research in setting driving hours regulations

Narelle Haworth

5.1 Driving hours regulations

Most jurisdictions throughout the world have regulations which limit driving hours for drivers of heavy vehicles. The regulations commonly include a limitation on the maximum number of hours that can be driven per day or per week and specifications relating to rest periods. Some jurisdictions limit actual driving hours (that is, hours behind the wheel) whereas others limit working hours (which may include loading, paperwork, waiting, etc.).

The findings of fatigue research can be used to design driving hours regulations to maximize road safety benefits, but road safety is only one of a number of issues affected by these regulations. In their Technical Working Paper No. 5, the National Road Transport Commission (1992: 1) acknowledged that 'the issue of driver working hours is contentious. The question raises trade-offs between productivity, road safety and occupational health objectives.' Restricting working hours is considered by many in the transport industry to reduce productivity, although this may not be so if the reductions are accompanied by improved occupational health. In addition, the enforcement of driving hours regulations aims to prevent undermining of competition by some hauliers demanding more of their drivers than others. Acknowledgment of these roles of driving hours regulations can help us to understand why driving hours regulations are so controversial and the subject of strong industrial and political pressure.

5.2 Driver working hours, fatigue and crashes

A number of studies have investigated the relationship between number of hours driven and crash rates.

In one of the earliest studies, Harris and Mackie (1972) found an increasing ratio of obtained to expected crashes between the seventh and tenth hours of driving. Harris (1977) showed that about twice as many crashes occurred in the second half of a driving trip, compared to the first. The crash rate was at its most inflated during the fifth and sixth hours and diminished somewhat thereafter. A study of French truck drivers (Hamelin, 1987), found that the crash rate increased dramatically after 12 or more hours of work.

Jones and Stein (1987) conducted a case-control study of large truck crashes in Washington State. For each large truck involved in a crash, three trucks were randomly selected from the traffic stream at the same time and place but one week later. These vehicles were inspected. The cases and controls were compared with respect to driver factors (age of driver, hours of driving, and logbook violations) and truck characteristics (carrier type and operation, truck load and fleet size). The factors which were associated with an increased risk of crash involvement were driving longer than eight hours, violation of logbook regulations, young drivers interstate carrier operations and equipment defects. Jones and Stein concluded that the crash risk for drivers of articulated vehicles who had driven for more than eight hours was double that of drivers who had driven for less than eight hours.

Analysis of crash statistics has also shown the importance of time-of-day factors in fatigue-related crashes. Overseas studies have shown that most crashes which result from drivers falling asleep at the wheel occur at night. In an analysis of US interstate truck crashes involving dozing drivers, Harris (1977) found that about twice as many of these crashes occurred between midnight and 8 a.m. as in the rest of the day and about half of the single-vehicle crashes occurred in the early morning hours. Studies conducted in France (Hamelin, 1987) and Sweden (Lisper et al., 1979) have demonstrated similar patterns of results. In an early study, Hoback (1959, cited in Lisper et al., 1979) concluded that 'a driver is 50 times more likely to go to sleep between 2 a.m. and 6 a.m. than from 8 a.m. to noon'. There is Australian evidence that more fatigue-related crashes occur at night than during the day. In a New South Wales study, Fell (1987) found that over 60 per cent of crashes judged by police officers to have involved driver fatigue occurred between midnight and 8 a.m. In another study, we (Haworth et al., 1989) analysed casualty and fatal truck crashes on five Victorian highways by time of day, controlling for truck volumes. Crash risks were highest during the night, suggesting the involvement of driver fatigue.

5.3 Implications of fatigue research for setting driving hours regulations

Analysis of crash data has taught us little about the contribution of fatigue to crashes that do not involve the driver falling asleep at the wheel. A general feeling of loss of attention, loss of interest and sometimes boredom accompanies the onset of fatigue. Loss of attention could result in taking a long time to brake, in choosing the wrong speed for a bend, in not noticing a vehicle entering the roadway. None of these crashes would be considered classic fatigue crashes, though.

One of the difficulties in examining the effects of fatigue, other than falling asleep at the wheel is the adaptability of drivers. Studies in the UK have shown that drivers who are affected by fatigue tend to compensate, to some extent, for these impairments by slowing down or being less willing to overtake (Brown et al., 1970). Naatanen and Summala (1978) concluded that 'research has not generally been able to show that driver fatigue increases the risk of an accident except by increasing the probability of falling asleep during driving'.

Fatigue research has shown that it is not just the number of hours worked that contributes to driver fatigue. Fuller (1984) concluded that 'the effects of prolonged driving depend in part on *when* that prolonged driving takes place rather than simply on its actual duration'. The results of our programme of experimentation support this conclusion. On the test track, most night-time subjects fell asleep before completing six hours of driving whereas only one daytime subject fell asleep. Departures from the driving lane increased with time at the wheel at night but not during the day. In the laboratory, night-time subjects more commonly fell asleep during six hours of a combined tracking and visual reaction time task. The frequencies of eye closures of longer than 500 msec duration increased much more at night (see Figure 5.1). Tracking errors also increased more with time on task in night-time than daytime testing.

Rest breaks

Another aspect of driver working hours regulations which has been the subject of exper-

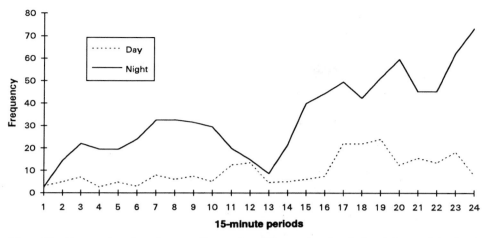

Figure 5.1 Frequencies of eye closures of longer than 500 msec duration during six hours of simulated driving during daytime or at night.

imental research is rest breaks. Experiments have examined how much benefit results from a rest break, when they should be scheduled, how long they should be and what should happen during a rest break.

The evidence for the effectiveness of breaks is mixed. It is unclear whether rest breaks of about one hour are more effective than short breaks of about 15 minutes. Some studies have even failed to demonstrate any beneficial effect of a rest break.

The mixed results which have been found may have resulted from when, during the trip, the break was taken. The results of several studies suggest that rest breaks may not be helpful once fatigue has developed. Harris and Mackie (1972) found that the truck drivers' first break after about three hours on the road produced physiological recovery and reduced driver errors but that later breaks after six and nine hours led to little improvement. Lisper *et al.* (1986) had subjects drive till they fell asleep and then take a five-minute walk. Subjects fell asleep again after an average of 23 minutes.

In one of our experiments, we gave subjects a 15-minute rest break after six hours of simulated driving (or after they had fallen asleep, if that happened first). In most cases, the improvement in steering, reaction time and eye closure resulting from the break disappeared within 30–45 minutes. Clearly, rest breaks have only a short-term benefit for drivers who are already very tired.

Rest breaks can be taken in the vehicle or outside, can involve exercise and can involve eating. It has not yet been established that exercise improves the effectiveness of a rest break. Several studies, including ours, have suggested that the intake of food may have a beneficial effect (Lauer and Suhr, 1959; Lisper and Eriksson, 1980).

The effect of prior activity

Our recent research has focused on the effect of prior activity on the development of driver fatigue. Examination of crashes had shown a number in which the driver had fallen asleep when driving after a day's work, sometimes after only a short period of driving. In our laboratory experiment, we found that drivers who had completed a day of physical work showed more evidence of fatigue in simulated driving than drivers who had completed a day of mental work or who had rested that day. When undertaking simulated driving after physical work, subjects spent significantly more time with their eyes closed and were less accurate in tracking. The effect of prior activity in hastening the onset of driver fatigue is not accounted for in the driving hours regulations.

5.4 Limitations of driving hours regulations

As mentioned earlier, driving hours regulations currently specify maximum hours driven (or worked) and rest periods. The regulations overlook the important factors of time of day of driving, activity during rest breaks, and prior activity which have been shown in experimental studies to affect the development of fatigue.

Currently, in Victoria at least, the driving hours regulations apply only to trips of greater than 80 km from the depot. Should the regulations apply for shorter trips? Very little is known about fatigue in short-distance driving, particularly in cities and towns. Whether this limitation on the regulations is valid cannot be determined currently from the fatigue research.

Much to the annoyance of many truck and bus drivers, driving hours regulations apply only to heavy vehicle drivers. They sometimes claim that similar regulations should apply to car drivers. Our study of fatigue in fatal crashes in Victoria involving trucks supports their claim (Haworth et al., 1989). The examination of coroners' reports showed that fatigue of car drivers was strongly involved in fatal truck crashes. The coroner identified car-driver fatigue in 9.2 per cent of truck crashes involving cars and fatigue on the part of articulated vehicle drivers in 5.1 per cent of crashes involving these vehicles. The difference in the estimates of fatigue between car drivers and articulated vehicle drivers is not statistically significant ($z = 1.2, p > 0.05$).

One might argue that driving hours regulations should be designed to prevent driving when driver impairment due to fatigue is of such a level that crash risk reaches a threshold value. Should driving hours regulations be based on concepts of degree of driver impairment, or degree of crash risk? Whether driver impairment or crash risk is chosen affects the way in which regulations should apply to driving in remote areas. In their discussion paper, the National Road Transport Commission (1992) has suggested different driving hours restrictions for populous and remote areas of Australia. It proposed less stringent restrictions on working hours and longer work cycles in remote areas, and a maximum number of working hours per year (see Table 5.1).

How does the proposal for different driving hours for trucks in the remote and the populous zones fit with what we know about driver working hours and fatigue and crashes?

Most people have experienced dozing off while driving long distances and suddenly finding themselves with a wheel in the gravel or over the centre line, or heading for a tree. Fortunately, in most of these instances the driver has managed to recover control without serious consequences. Thus, fatigue can sometimes be quite severe with no crash result. Crashes are often the combination of characteristics of the driver and characteristics of the driving environment.

The driving conditions result in a level of impairment of performance of the driver. Depending on the driving environment, a crash may or may not occur. Let us apply this model to driving in populous and remote areas. In comparison with populous areas, remote areas are likely to have fewer intersections, other vehicles, pedestrians, road-side poles and roadside trees.

Because of the longer trips and because of factors such as the fewer number of other vehicles, driving in a remote area is more monotonous and so is likely to lead to higher levels of driver fatigue. The probability of dozing off, of leaving the roadway or drifting

Table 5.1 Restrictions on driver working hours suggested by National Road Transport Commission

		Trucks	
Attributes	Buses	populous zone	remote zone
Maximum hours per 24 hours per driver	14	14	ns
Length of working cycle in days	7	7	14
Maximum continuous 24 hours rest period per cycle	1	1	3
Minimum continuous hours of rest per 24 hours	5	5	5
Maximum period of continuous working in hours	5	5	5
Minimum number of minutes of rest for each maximum period of continuous work	30	30	30
Minimum recognized rest period in minutes	15	15	15
Maximum working hours per year per driver	3500	3500	4000

across the road is higher in these areas. The probability of having a crash would be much higher if it was not for the lack of other vehicles, poles, etc. Thus in remote areas, we have the strange phenomenon that the very characteristics which increase fatigue, reduce the risk of impaired driving resulting in a crash.

5.5   An alternative to driving hours regulations

It is possible that driving hours regulations which were valid in terms of minimizing driver fatigue would be very complex and therefore difficult to understand and enforce and unworkable in practical terms. Thus the choice may be to retain a suboptimal set of driving hours regulations or to change to an alternative method of reducing driver fatigue levels.

Systems for the detection of driver fatigue can be considered as a possible future alternative to driver working hours regulations, at least to the extent to which such regulations are designed to ensure safety.

A system for detection of driver fatigue comprises a measure of fatigue, a standard against which the value of the measure is compared and a mechanism of communicating a finding that performance is degraded (Haworth, 1992). Systems need to be reliable, have an appropriate degree of sensitivity and not be intrusive to the driver.

Systems have taken two forms: performance tests (administered before work or at the roadside) and in-vehicle systems. Performance tests have been developed and are being tested in the US (Allen et al., 1990). Tests administered before work are most suited for use by individual companies. For general enforcement purposes, however, a system which does not require such data is needed.

We have evaluated a range of commercially available in-vehicle fatigue monitors. In general, the devices showed the ability to detect fatigue in some cases but either did not have the appropriate degree of sensitivity or were intrusive to the driver.

We are currently working on validation of an in-vehicle system for detection of driver fatigue. This system analyses small steering corrections. Alert drivers frequently make such corrections but steering movements become less frequent and larger as drivers become tired. This system has the advantage of not being at all intrusive to the driver.

Overseas research has suggested that it will detect the effects of alcohol as well as those of fatigue.

5.6 Conclusions

Driving hours regulations are a legislative approach to reducing the contribution of fatigue to heavy vehicle crashes. The omission of important factors affecting alertness levels, such as time of day, activity during rest breaks and prior activity, means that these regulations are incapable of being completely effective, even if problems related to enforcement were solved.

In truck driver fatigue, driver working hours are not the most important issue. The most important issue is the degree of alertness of the driver at any point in driving, whether he has just left the depot or whether he has already driven for twelve hours. What road safety researchers and practitioners would like to do is ensure that drivers are not impaired, that they are sufficiently alert to be able to respond to the demands of the road and other road users. From a road safety point of view, driving hours are a measure of impairment but a poor one.

Acknowledgment

The financial support provided by the Federal Office of Road Safety and VIC ROADS is gratefully acknowledged.

References

Allen, R.W., Stein, A.C. and Miller, J.C., 1990, Performance testing as a determinant of fitness-for-duty, Paper 457, Presented at SAE Aerotech '90, Long Beach, California, Oct 1–4, 1990.

Brown, I.D., Tickner, A.H. and Simmonds, D.C.V., 1970, Effect of prolonged driving on overtaking criteria, *Ergonomics*, **13**, 239–42.

Fell, D., 1987, A new view of driver fatigue, Draft version, Rosebery, NSW: Traffic Authority of NSW.

Fuller, R.G.S., 1984, Prolonged driving in convoy: the truck driver's experience, *Accident Analysis and Prevention*, **16**, 371–82.

Hamelin, P., 1987, Truck driver's involvement in traffic accidents as related to their shiftworks and professional features, Symposium on the role of heavy freight vehicles in traffic accidents, Montreal, Canada, pp. 3:107–30.

Harris, W., 1977, Fatigue, circadian rhythm, and traffic accidents, in Mackie, R.R. (Ed.), *Vigilance: Theory, Operational Performance, and Physiological Correlates*, New York: Plenum, pp. 133–46.

Harris, W. and Mackie, R.R., 1972, A study of the relationships among fatigue, hours of service, and safety of operations of truck and bus drivers, Report No. BMCS-RD-71-2. Washington, DC: Bureau of Motor Carrier Safety.

Haworth, N.L., 1992, Systems for detection of driver fatigue, *Proceedings 16th ARRB Conference*, **Part 4**, 189–201.

Haworth, N.L., Heffernan, C.J. and Horne, E.J., 1989, Fatigue in truck accidents, Report No. 3, Melbourne: Monash University Accident Research Centre.

Jones, I.S. and Stein, H.S., 1987, *Effect of driver hours of service on tractor-trailer crash involvement*. Washington, DC: Insurance Institute for Highway Safety.

Lauer, A.R. and Suhr, V.W., 1959, The effect of a rest pause on driving efficiency, *Perceptual and Motor Skills*, **9**, 363–71.

Lisper, H.O. and Eriksson, B., 1980, Effects of length of a rest break and food intake on subsidiary reaction-time performance in an 8-hour driving task, *Journal of Applied Psychology*, **65**, 117–22.

Lisper, H.O., Laurell, H. and van Loon, J., 1986, Relation between time to falling asleep behind the wheel on a closed track and changes in subsidiary reaction time during prolonged driving on a closed track, *Ergonomics*, **29**, 445–53.

Lisper, H.O., Eriksson, B., Fagerstrom, K.O. and Lindholm, J., 1979, Diurnal variation in subsidiary reaction time in a long-term driving task, *Accident Analysis and Prevention*, **11**, 1–5.

Naatanen, R. and Summala, H., 1978, Fatigue in driving and accidents, in *Driver fatigue in road traffic accidents*, Report EUR 6065 EN, Luxembourg: Commission of the European Communities.

National Road Transport Commission, 1992, Driver working hours, *Technical Working Paper No 5*, Melbourne: NRTC.

SECTION TWO:
The epidemiology of fatigue-related crashes

6 Road traffic crashes by region in Western Australia

G. Anthony Ryan

6.1 Introduction

About 200 people die and about 13 000 are injured in road traffic crashes reported to the police in Western Australia each year. To examine in more detail the patterns of these road traffic crashes, Western Australia can be conveniently subdivided into three regions (Figure 6.1):

- the Perth statistical division;
- the Rural region made up of the statistical divisions of Midlands, Upper Great Southern, Lower Great Southern, and South West; and
- the remote region, consisting of the Central, Pilbara, Kimberley and South East statistical divisions.

These regions roughly correspond to the metropolitan, agricultural, and the mining and pastoral areas of the state.

The Road Injury Database of the Road Accident Prevention Research Unit was used to examine the characteristics of crashes in each region. This database consists of records of all police reported casualty crashes, all hospital discharges and all deaths from road traffic crashes in Western Australia for 1988.

6.2 Results

Just over 70 per cent of the population of Western Australia live in the Perth region, with 16 per cent and 12 per cent in the rural and remote regions respectively (Table 6.1). The population of the remote region is somewhat younger than that of the other regions. The Aboriginal population forms about 1 per cent of the population of the Perth region and 12 per cent of that of the remote region, and is markedly younger than the non-aboriginal population, with about 40 per cent aged less than 15 years.

Table 6.2 shows that while just under half of the deaths and three quarters of the injuries occur in the Perth region, crashes in the rural and remote regions tend to be more lethal, with about 3 per cent and 5 per cent respectively of casualties dying compared with 1 per cent for the Perth region.

Patterns of crash

The types of crash occurring in each region have characteristic patterns (Table 6.3). There are relatively more crashes on weekends and at night in the rural and remote regions, compared with Perth. Perth has more rear-end and angle collisions, but fewer off-road crashes and collisions involving animals. Perth also has more crashes at intersections and on wet roads, while the rural and remote regions have more crashes on curves and unsealed roads. The remote region has relatively more crashes involving motorcycles.

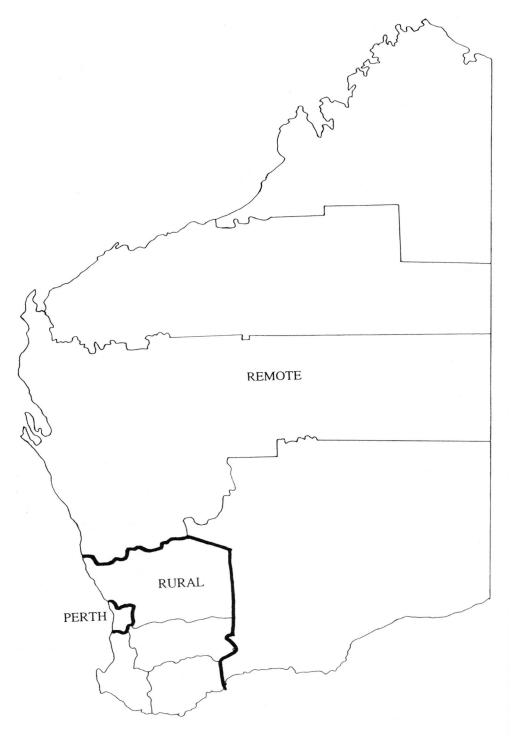

Figure 6.1 Map of Western Australia showing the Perth, rural and remote regions, with the boundaries of the statistical divisions within each region. Western Australia is 2300 km from north to south and 1600 km from east to west.

Table 6.1 Age, race and region

	Age (%)				Total	Aboriginal %
	0–14	15–24	25–59	60+		
Region						
Perth	22.2	17.2	46.7	13.8	1,118,772	0.9
Rural	25.4	14.8	44.9	14.9	244,078	2.2
Remote	27.0	17.2	48.4	7.4	181,956	12.3
Race						
Aboriginal	39.4	23.3	32.0	5.2	37,788	
Non-aboriginal	23.1	16.7	47.0	13.4	1,507,018	

Source: Population, WA 1988.

Table 6.2 Deaths and casualties by region

	Deaths		Casualties		Total	Percentage deaths
	n	%	n	%	n	
Region						
Perth	110	46.2	10617	77.1	10727	1.0
Rural	60	25.2	1800	13.1	1860	3.2
Remote	68	28.6	1353	9.8	1421	4.8
Total	238	100.0	13770	100.0	14008	1.7

Source: Police reports, WA, 1988.

Crashes involving casualties in the Perth region are characterized by a large proportion of rear and angle collisions, with fewer single-vehicle crashes (Table 6.4). The majority of pedal cycle and pedestrian casualties also occur in the Perth region. The largest numbers of deaths occur to pedestrians, and in single-car crashes, in single-vehicle motorcycle crashes and angle car crashes. In the rural region, the largest group of both deaths and casualties is single-car crashes. Rear and angle car collisions were the next most common sources of casualties. A similar pattern is also found in the remote region, with single-vehicle car, truck and motorcycle crashes being relatively very frequent. Single-vehicle and head-on car crashes are the most frequent cause of deaths in this region. It is worth noting that the proportion of trucks involved in single-vehicle crashes increases with increasing distance from Perth. The highest percentage of deaths is found in single-vehicle crashes and in head-on car crashes in the remote region.

Place of residence

While over 75 per cent of casualties in each region occur to residents of the region, 20 per cent and 11 per cent respectively of casualties in the rural and remote regions were residents of Perth (Table 6.5). Therefore, while there may be a kernel of truth in the belief that city residents are disproportionately involved in rural crashes, it is still true that the majority of rural crashes involve local residents.

Age and sex

The rate of hospital admission from road traffic crashes increases with increasing age

Table 6.3 Road traffic crashes by region

Variable	Perth %	Rural %	Remote %
Day of week			
Fri/Sat/Sun	42.3	47.9	45.8
Time of day			
6 p.m.–midnight	20.6	25.3	27.2
midnight–6 a.m.	5.2	7.5	11.1
Nature of crash			
Rear end	37.5	11.0	5.7
Opposite direction	3.8	6.8	7.8
Side swipe, same direction	5.9	5.9	3.8
Right angle	33.0	15.9	7.3
Pedestrian	5.4	2.7	6.5
Animal	0.2	1.3	3.0
Off road	11.2	54.3	59.8
Site			
Intersection	60.3	24.2	19.2
Curve	14.9	30.4	31.2
Wet	26.3	17.3	9.5
Unsealed	0.6	17.7	25.4
Vehicle			
Car	88.1	88.3	79.2
Truck	3.6	4.2	6.1
Motorcycle	8.3	7.4	14.7
Number of crashes	11337	1926	1408

Source: Police reports, WA, 1988.

Table 6.4 Deaths, casualties, crash type and region

	Perth			Rural			Remote		
Crash type	Deaths	Casualties	%	Deaths	Casualties	%	Deaths	Casualties	%
Car vs car									
Head-on	8	301	2.6	5	92	5.1	14	77	15.4
Rear	5	4001	0.1	2	220	0.9	1	64	1.5
Angle	10	2918	0.3	2	195	1.0	1	100	1.0
Single car	25	1000	2.4	38	944	3.9	35	715	4.7
Truck									
Multiple	6	404	0.2	2	63	3.1	4	46	8.0
Single	1	26	3.7	6	39	13.3	2	56	3.4
Motorcycle									
Multiple	8	564	1.4	1	58	1.7	0	81	0.0
Single	13	209	5.9	3	46	6.1	3	62	4.6
Pedal cycle	6	599	1.0	1	79	1.2	1	44	2.2
Pedestrian	28	584	4.6	0	50	0.0	7	91	7.1

Source: Police reports, WA, 1988.

Table 6.5 Place of residence and region

Region of crash	Perth (%)	Region (%)
Perth	87.6	87.6
Rural	20.1	74.4
Remote	11.4	79.2

Source: Hospital admissions, road traffic crashes, WA, 1988

Table 6.6 Region and age and sex

	Age (years)							Sex	Rate/100,000 population
	0–4	5–9	10–14	15–19	20–24	25–59	60+		
Perth	93	196	261	576	508	218	200	Male	348
								Female	179
Rural	163	203	227	931	768	244	168	Male	366
								Female	252
Remote	107	264	351	838	948	380	142	Male	566
								Female	228

Source: Hospital admissions, road traffic crashes, WA, 1988 (rate per 100,000 population).

Table 6.7 Road user type and region

	Driver (%)	Passenger (%)	Motor cycle (%)	Pedal cycle (%)	Pedestrian (%)	Not known (%)	Total (n)
Perth	19.4	17.1	15.4	20.0	11.3	16.8	2671
Rural	14.2	15.6	12.4	11.7	3.3	42.8	733
Remote	10.1	19.3	17.7	8.9	6.9	36.8	734

Source: Hospital admissions, WA, 1988.

to a maximum at 15–19 years in the Perth and rural regions and 20–24 years in the remote region (Table 6.6). At each age group, Perth tends to have the lowest rate, with the remote region the highest, and the rural region an intermediate value. The exception is the 60+ age group where the highest rate is found in Perth and the lowest in the remote region. Within each region, the male rate is about twice that of the female rate.

Type of road user
From the hospital admission records it is possible to determine the type of road user involved. The percentage of cases in which the road user type is unknown is much larger in the rural and remote regions than in Perth (Table 6.7). The largest single group of admissions in Perth is pedal cyclists, followed by drivers and passengers. In the rural region, the largest groups are passengers and drivers, while in the remote region the largest groups are motorcyclists and passengers.

Table 6.8 Age, road user type and region

	Perth			Rural			Remote		
Age (years)	0–4 (%)	5–9 (%)	10–14 (%)	0–4 (%)	5–9 (%)	10–14 (%)	0–4 (%)	5–9 (%)	10–14 (%)
Passenger	31.6	12.3	11.2	51.4	43.9	30.2	57.9	41.9	45.3
Pedal cycle	47.4	65.2	67.4	28.6	43.9	58.1	15.8	39.5	35.9
Pedestrian	17.1	17.4	15.8	17.1	9.8	4.6	21.0	16.3	5.7

Source: Hospital admissions less than 15 years of age, WA, 1988.

Type of road user and age

For those aged less than 15 years, there are different patterns of involvement in each region (Table 6.8). In Perth, for each of the age groups 0–4, 5–9 and 10–14 years, the largest group is involved as pedal cyclists. In the rural region, the 0–4 year group is involved as passengers, the 5–9 year group as both passenger and pedal cyclists, and the 10–14 year group as pedal cyclists and to a lesser extent as passengers. In the remote region, the primary method of involvement at each age group is as passenger. In each region, the frequency of involvement as a pedal cyclist increases from the age group 5–9 years. Involvement as a pedestrian is relatively low in each region.

Aboriginals

The rate of admission of Aboriginals is almost three times higher than for non-Aboriginals, for all age groups except those over 60 years (Table 6.9). The highest rate, found in 15–19 year olds, is 1209 admissions per 100 000 population. This means that more than 1 in 100 Aboriginals of this age will be admitted to hospital each year due to a road traffic crash. Unlike non-Aboriginals, the rate for the Perth region is intermediate between the rural and remote regions.

Aboriginals have different patterns of involvement as road users, compared with non-Aboriginals (Table 6.10). In Perth, Aboriginals are involved as pedestrians, passengers and pedal cyclists. In the rural region, they are involved as passengers and pedal cyclists and, in the remote region, as passengers and as pedestrians.

6.3 Discussion and summary

About 46 per cent of the deaths and 77 per cent of the casualties from road traffic crashes occur in the Perth region. Crashes occurring in the rural and remote regions are more lethal, with a higher percentage of casualties dying than in the Perth region.

Table 6.9 Age, race and region

	Age (years)								Region		
	0–4	5–9	10–14	15–19	20–24	25–59	60+	all ages	Perth	Rural	Remote
Aboriginal	280.5	574.1	493.2	1209.0	1074.0	834.6	102.1	714.5	624.6	521.4	802.6
Non-Aboriginal	99.7	191.0	256.8	634.2	584.3	231.2	191.5	277.6	260.0	305.8	357.0

Source: Admissions, road traffic crashes, WA, 1988 (rate per 100,000 population).

Table 6.10 Race, road user type and region

	Driver (%)	Passenger (%)	Motorcycle (%)	Pedal cycle (%)	Pedestrian (%)	Not known (%)	Total (n)
Perth							
Aboriginal	11.3	17.7	0.0	17.7	35.5	17.7	62
Non-Aboriginal	19.6	17.1	15.8	20.1	10.7	16.6	2609
Rural							
Aboriginal	0.0	23.1	0.0	15.4	11.5	50.0	26
Non-Aboriginal	14.7	15.3	12.9	11.6	3.0	42.6	707
Remote							
Aboriginal	3.3	27.5	5.6	2.8	10.7	50.0	178
Non-Aboriginal	12.2	16.7	21.6	11.0	5.8	32.7	556

Source: Hospital admissions, WA, 1988.

These differences are associated with the different patterns of crash occurring in each region. In Perth, there is a high proportion of rear-end and angle collisions, while in the other regions there is a much higher proportion of single-vehicle crashes, which tend to have more fatalities. Crashes involving pedestrians and pedal cyclists are concentrated in the Perth region. Trucks are involved in multiple-vehicle rear-end crashes in Perth, and in single-vehicle crashes in the other regions. Pedal cycle crashes are the most frequent cause of hospital admission in Perth, and for those aged less than 15 years in all regions. The age group 5–15 years is most frequently involved as a passenger in a motor vehicle. Rather surprisingly, motorcyclists are frequently involved in single-vehicle crashes particularly in the remote region. Aboriginals have a much higher rate of admission than non-Aboriginals, being involved primarily as passengers and as pedestrians. The hospital admission rate is highest in the remote region, followed by the rural and Perth regions. This progression parallels the consumption of alcohol, as shown by a recent publication which shows that the Kimberley, Pilbara and South East regions have a per capita consumption of alcohol twice that of the rest of the state (Philp and Daly, 1993).

It is only possible to speculate on the causes for the patterns and differences found in this survey. The high fatality rate in the remote and rural regions may be due to higher travelling speeds, lower rates of seat-belt wearing and higher alcohol consumption. There may also be delays in the provision of medical care associated with long distances between towns and sparse populations. The distinctive patterns of involvement of Aboriginals may be due to the manner in which motor vehicles are used, with many passengers often using the one vehicle. Further research is planned in order to determine the answers to these problems.

References

Philp, A. and Daly, A., 1993, Alcohol consumption in selected regions in Western Australia, July 1989 to June 1991, Technical Report, Perth: Western Australian Alcohol and Drug Authority.

7 Fatal accidents in Western Australia

Peter J. Moses

7.1 Introduction

Osmar White, in his *Guide to Australia* (1968), said of touring in Western Australia:

> The voyager in the west must be prepared to take sightseeing in Homeric doses, to travel hundreds of miles through a monotony of dreary mulga, saltbush or spinifex to experience the delight of discovering some scenic oasis – perhaps some deep, green waterhole in a dried-up river bed ringed with pandanus jungle and teeming with bird and animal life, or a procession of peaks or table topped mountains fantastically weathered and brightly stained by leached minerals.

Although this was written 25 years ago, little has changed; the distances are vast and accidents happen, sometimes unfortunately with fatal results. Detailed records are kept for fatal accidents and this chapter draws on these records to identify some common causes.

Over the last 20 years, the number of fatalities on Western Australia's roads has decreased. This decrease is better understood if one views the fatal accident scene in the form of a rate such as fatalities per 10 000 vehicles registered, 100 000 population or billions of kilometres travelled.

7.2 Accident rates

The fatal accident statistics are examined in four ways:

1. **Gross fatalities recorded**
   The average gross fatality has decreased from approximately 350 in 1972/73, to approximately 200 in 1990/91, a reduction of 43 per cent.

2. **Fatalities per 10 000 vehicles registered**
   The average fatalities per 100 000 vehicles registered has decreased from approximately 7.3 in 1972/73, to approximately 1.9 in 1990/91, a reduction of 74 per cent.

3. **Fatalities per 100 000 population**
   The average fatalities per 100 000 population has decreased from approximately 32.3 in 1972/73, to approximately 12 in 1990/91, a reduction of 63 per cent.

4. **Fatalities per billion kilometres travelled**
   The average fatalities per billion kilometres travelled has decreased from approximately 48.3 in 1970/71, to approximately 15.5 in 1988/89, a reduction of 68 per cent.

7.3 Rural fatal accidents

In this analysis, the cause and circumstances of these rural fatal accidents are examined for all reported non-metropolitan fatal accidents for a calendar year. The accident chronology has been mixed to ensure confidentiality.

- Areas examined
  - The involvement of rural vs city-based drivers is examined
  - The age of the driver of the vehicle
  - Conditions of light or darkness
  - Whether the accident was a single- or multiple-vehicle event
  - The stated main cause of the accident
- Accident base

In all, 86 accidents were examined with data being drawn from police accident reports.

7.4 Evaluation

Origin of driver
For this analysis drivers who were from outside the state were classified as city drivers. As shown in Figure 7.1 the predominant driver involved in rural fatal accidents for the year came from non-metropolitan addresses.

Age of the driver
Of the 84 accidents where an age of the motorist was indicated, Figure 7.2 shows the age dispersion of the motorists. In the case of multiple accidents, the age of the motorist apparently more at fault is used. It is apparent that 50 per cent of the motorists involved in rural fatal accidents were 25 years of age or less. Therefore it would be prudent to target the young driver, particularly from rural areas, with an educational programme of countermeasures to fatigue and to provide driving techniques on gravel roads.

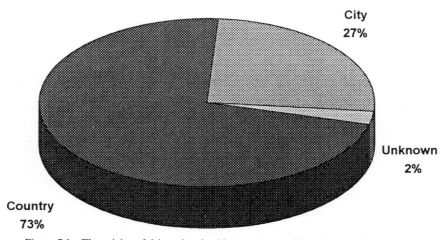

Figure 7.1 The origins of drivers involved in non-metropolitan accidents for one year.

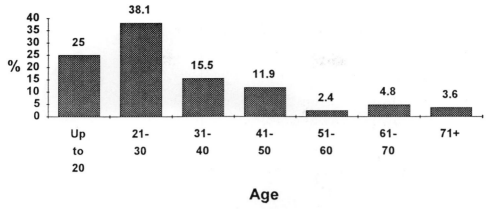

Figure 7.2   The age of drivers involved in non-metropolitan accidents for one year.

Light conditions

As shown in Figure 7.3, of the 86 accidents examined, 47 (or 54.7 per cent) occurred in daylight and 39 (or 45.3 per cent) occurred at night-time. Although rural roads carry only approximately 15–20 per cent of their daily traffic between the night-time hours of 7 p.m. and 7 a.m., almost half of the rural fatal accidents occurred at night. The need for good night-time road delineation is demonstrated here and dealt with later in the chapter.

Vehicle involvement

In this analysis, where other than a single vehicle is involved, such as in the case of vehicle colliding with a pedestrian, the accident is considered to be a multiple one. As shown in Figure 7.4, of the 66 accidents available for analysis, 60 (or 69.8 per cent) concerned single vehicles; 26 (or 30.2 per cent) involved multiple vehicles. Rural fatal accidents are predominantly single-vehicle accidents.

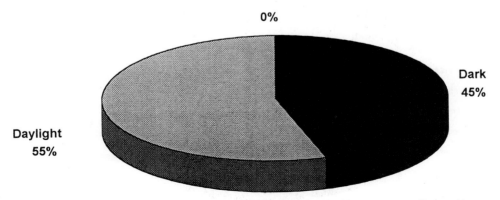

Figure 7.3   The lighting conditions which prevailed for drivers involved in non-metropolitan accidents for one year.

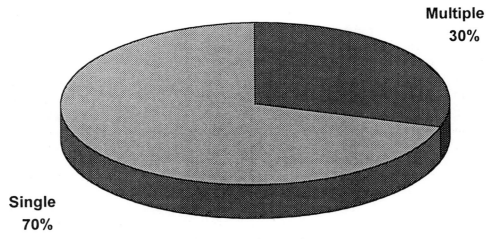

Figure 7.4 Single- or multiple-vehicle involvement in non-metropolitan accidents for one year.

7.5 Main pattern of accidents

As shown in Figure 7.5, it is clear that by far the major pattern of fatal accidents in rural areas is that the vehicle left the sealed road, overcorrected having struck the gravel shoulder and either overturned or hit roadside objects, normally trees. Such occurrences indicate that at least inattention and inexperience were involved in these accidents and while going to sleep or fatigue is not nominated by the police list as a cause of the accident, it is clear that such a supposition would be sustainable.

7.6 Some countermeasures

Countermeasures to combat fatigue are varied and include publicity to make drivers aware of the dangers of fatigue, the installation of visual or audio stimulation within

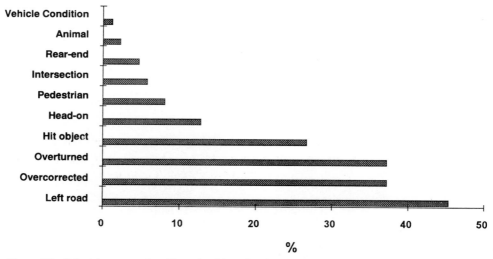

Figure 7.5 Principle causes of accidents for drivers involved in non-metropolitan accidents for one year.

the road environment and the provision of suitable rest areas so that fatigued drivers can rest before resuming their journey.

Visual stimulation

The level of visual stimulation one receives from the environment in which one is travelling varies greatly with the individual. It may in fact change over a period of time where subtleties which were not seen initially become perceived as interesting when the environment is searched more carefully, or it can be man-made stimuli installed into a landscape in the form of kilometre pegs, signs, road design or beautification. Some of the most tedious in roads in Western Australia are straight, disappearing into the distance without deviation in line. By the inclusion of curves in the initial design, the driver is afforded increased visual stimulation.

Items such as kilometre pegs, which stimulate a search for the next peg, followed by mental calculations of the portion of the journey completed, can also enhance driver alertness when compared to a similar section not so treated. There also seems to be a great area to be exploited by landscape specialists in providing interesting areas of beautification and plantings, not only visually to intrigue the driver but also to allow for areas of rest protected from the elements.

By careful consideration, the road designer, landscape architect and traffic engineer should be able to increase the level of visual stimulation of given sections of road and thereby the inherent level of safety.

Painted edge lines

In the mid 1980s, the length of edge lines painted on major roads in the State increased dramatically. On one occasion, over 650 km of edge line was emplaced in one year on major roads radiating from Perth. As the State reported damage-only accidents as well as casualties, it was possible to examine the effect of these lines on accident occurrence.

Roads treated

The roads treated with edge lines were:

- Albany Highway – 112 km
- Brand Highway – 266 km
- Great Eastern Highway – 164 km
- South Western Highway – 110 km

Traffic volumes

Traffic volumes on the roads considered were increasing ranging from 1.9 per cent on South Western Highway to 3.8 per cent on Brand Highway, so estimates of the benefit of edge lining obtained by comparing accident rates the year before and after emplacement are considered conservative.

Alcohol and vision

It is generally accepted that 90 per cent of all information for the driver on our roads comes normally through vision. The effect of alcohol on this sense is therefore critical.

It is accepted that alcohol causes a four-fold deterioration:

- Alcohol reduces the sensitivity to visual contrast which has specific application in relation to retroreflective signs and pavement markings which are based on visual contrast for their effectiveness.
- Alcohol reduces peripheral vision, thus information will tend to come from the centre of the road.
- It induces tunnel vision which makes the driver concentrate on objects close to the front of the vehicle.
- The alcohol-affected driver cannot process information as well as the sober driver.

Studies in the US have shown that by using a wider edge line the visual perception of edge lines by those affected by alcohol was increased, resulting in better vehicle trajectories on roads treated with wider edge lines.

Type of edge line

As a result of accepting the findings of the above study, the edge lines painted on the four highways radiating from Perth were 150 mm in width, the night-time impact being enhanced by impregnating the line with glass beads. Whilst the Australian Standard only required a 80 mm wide line, with a preferred width of 120 mm, Main Roads chose to install a wider edge line as a positive measure to reduce single-vehicle accidents involving the alcohol-affected or tired driver.

Results

Total accidents

The results of the installation of these lines for the 652 km of road considered on a year-before and year-after basis are shown in Table 7.1. The fall from 83 to 55 reported 'out of control' accidents shows a reduction in occurrence of this type of accident to 66 per cent of its original total.

Single-vehicle accidents

Analysing the single-vehicle accidents, the most significant result was the halving of run-off road accidents not involving alcohol or mechanical failure as shown in Table 7.2.

Table 7.1 Total accidents and single-vehicle accidents before and after edge line emplacement

Highway	Length km	Total accidents		Out of control		Percentage of total	
		Before	After	Before	After	Before	After
Albany	112	49	48	15	12	31	25
Brand	266	70	67	19	15	27	22
Great Eastern	164	91	79	33	17	36	22
Southwest	110	60	55	16	11	27	20
Total	652	270	249	83	55	31	22

Table 7.2 Single-vehicle accidents before and after edge line emplacement

Highway	Out of control		Alcohol		Mechanical failure		Other	
	Before	After	Before	After	Before	After	Before	After
Albany	15	12	3	2	5	6	7	4
Brand	19	15	5	2	7	9	7	4
Great Eastern	33	17	6	6	3	1	24	10
Southwest	16	11	3	3	1	1	12	7
Total	83	55	17	13	16	17	50	25

The incidence of alcohol-affected accidents declined to about 75 per cent of its original level while the number of accidents involving mechanical failure such as blow outs quite logically remained the same.

Head-on accidents

With edge lines emplaced, it could be surmized that head-on accidents would increase. However, as Table 7.3 shows there was no change to the occurrence of this category of accident.

Other benefits

The level of shoulder maintenance subsequent to the installation of edge lines is dramatically reduced, effecting very significant savings in that area. Edge lines also assist vulnerable road users such as cyclists, allowing them to travel on a sealed shoulder with enhanced safety.

Profile edge lines

A more recent development has been the installation of profile edge lines which combine the visual advantages of painted edge lines with an audible noise when driven over. Profiling also enhances the visibility of lines by forming a series of vertical reflecting surfaces. Some experimental lengths have been emplaced in Western Australia. While guiding the errant motorist, this device should not be used within 200 m of a residence because, under low ambient noise conditions, residents are disturbed by the noise.

Table 7.3 Head-on accidents before and after edge line emplacement

Highway	Head-on	
	Before	After
Albany	4	3
Brand	5	2
Great Eastern	6	8
Southwest	3	4
Total	18	17

Rumble strips

Another area experimented with over the last two or three years has been rumble strips, a series of sequential thermoplastic strips placed across a carriageway to alert the motorist to a change of environment such as approaching a rural town site or isolated school. This has the same restrictions in use as the profile edge lines.

Focal point markers

These markers, set at 5 or 10 km intervals, add interest and information for drivers allowing them to gauge their journeys.

7.7   Conclusion

Currently, in this state, we are at the start of studying fatigue and introducing countermeasures and can learn much from others who are further down the long road which will lead to an understanding of this complicated area of traffic safety. We have a strong vested interest in combating what is a real problem, considering the vastness of our state.

Further reading

Maisey, G., 1992, *Road crashes in WA 1991 –. The Statistics*, WA Police Department.
Moses, P., 1986, *Edge lines and Single Vehicle Accidents*, Main Roads WA.
Moses, P., 1989, *Road Safety Techniques in Europe and North America*, Main Roads WA.
Moses, P., 1991, *Things that go bump in the night*, Main Roads WA.
White, O., 1968, *Guide to Australia*, Melbourne: Heinemann.

8 Drugs, driving and enforcement

Gavin Maisey*

8.1 Background

The influence of alcohol in road crashes, particularly those involving serious injury, is well documented (Lloyd, 1992; West and Hore, 1980). In Western Australia, for example, over one-quarter (27 per cent) of fatal road crashes in 1992 involved a driver with a blood-alcohol concentration (BAC) at or over the previous legal blood-alcohol limit of 0.08 per cent for drivers (WA Police Department, 1993). This proportion is much less than that recorded ten years ago (37 per cent in 1982). The number of such crashes has also fallen from over 100 in 1980 to 46 in 1992. Thus, while there has been a general downward trend in fatal crashes, this trend has been more pronounced for those crashes involving intoxicated drivers.

Much of the reduction in serious alcohol-related crashes may be attributed to the success of random breath testing (RBT) and earlier programmes of random stopping. In the random stopping programmes, motorists were stopped for checks of driver licence information and motor vehicle road worthiness. Drivers impaired by alcohol or other drugs were often detected in such police checks. This *de facto* RBT in Western Australia was replaced by a formal RBT law on 1 October 1988.

The success of RBT programmes in other Australian jurisdictions (Homel, 1986; Homel *et al.*, 1988) had provided considerable impetus for the development of a similar programme in Western Australia. A number of evaluations of the RBT programme in Western Australia have since been conducted (Traffic Board of Western Australia, 1989; Traffic Board of Western Australia, 1993). It has been shown to be a most efficient method for stopping and testing motorists for alcohol, and effective in deterring motorists from driving while impaired by alcohol and thus reducing the incidence of crashes involving intoxicated drivers. Stockwell *et al.* (1991) suggested a number of changes to enhance the local programme (e.g. increasing the proportion of stopped drivers who are tested for alcohol) and monitoring and refinement is continuing.

One strategy has involved the collection of information on the drinking location of drivers arrested by police following a road crash, or detected in police operations such as RBT. This programme, conducted in conjunction with the National Centre for Research into the Prevention of Drug Abuse, has enabled police to target drink–driving operations better. Lang and Stockwell (1991), for example, found that unlicensed locations such as private residences or public places (e.g. parks) were more likely than licensed premises to be the site of drinking for those subsequently involved in road accidents.

Limited information is available on drugs and driving in Western Australia. From data collected by the Chemistry Centre, however, Campbell (1990) reports an increasing involvement of cannabis and other illicit drugs by drivers involved in serious crashes.

The views expressed are those of the author and do not necessarily reflect those of the WA Police

Table 8.1 Drugs associated with traffic deaths in Western Australia 1990/91

Drug type	Total detected	Impaired
Cannabis	18	12
Benzodiazepines	9	0
Opiates	9	0
Stimulants M	7	6
Barbiturates	1	0
Other narcotics	6	?
Other hypnotics	1	1
All other drugs	43	0

Source: Campbell, Chemistry Centre WA.

Table 8.1 shows an update of this data: the number of drugs associated with traffic deaths in 1990/91, and an assessment of the number of cases where impairment was likely to have occurred based on case details and concentrations of drugs.

In a study of high-speed police pursuits in Perth, Homel (1990) found that charges for driving under the influence of alcohol and/or drugs were relatively low, but that this may have been due to factors associated with the process of obtaining a conviction for reckless driving. Only 23 per cent of offenders were charged with a drink–driving offence and the proportion of offenders charged with drug offences was negligible. Yet in interviews with offenders, it was found that most confessed to drug (e.g. amphetamines) or alcohol impairment at the time of the pursuit.

Table 8.2 Penalties for driving under the influence of alcohol and/or drugs

OFFENCE CATEGORY (BAC)	Offence type		
	1st	2nd	Subsequent
0.02%+	$100–300 or CSO *and* 3 mths or longer loss of MDL	$100–300 or CSO *and* 3 mths or longer loss of MDL	$100–300 or CSO *and* 3 mths or longer loss of MDL
0.05–0.079%	$100 *and* 3 demerit points	$100 *and* 3 demerit points	$100 *and* 3 demerit points
0.08–0.149%	$300–800 or CSO *and* 3 mths or longer loss of MDL	$600–1200 or CSO *and* 6 mths or longer loss of MDL	$600–1200 or CSO *and* 6 mths or longer loss of MDL
0.15%+ *plus* DUI *plus* refusing a test	$500–1200 or CSO *and* 6 mths or longer loss of MDL	$1000–1800 or 6 mths imprisonment *or* CSO *and* 2 yrs or longer loss of MDL	$1200–2500 or 18 mths imprisonment *and* loss of MDL for life

- CSO denotes Community Service Order
- MDL denotes motor drivers licence
- DUI denotes driving under the influence of alcohol or drugs
- 0.02% offence is only for first-year (probationary) drivers

8.2 Enforcement

Alcohol and drug driving limits

A number of categories of *per se* offences and penalties exist for driving while impaired by alcohol and/or other drugs. These include driving:

- with a blood–alcohol concentration (BAC) of 0.02 per cent or higher (for a learner or first year (probationary) driver);
- with a BAC of 0.05 per cent and less than 0.08 per cent (introduced 16 June 1993);
- with a BAC of 0.08 per cent and less than 0.15 per cent;
- with a BAC of 0.15 per cent or higher; and
- under the influence of alcohol and/or drugs.

Table 8.2 shows the penalties for various offences of driving under the influence of alcohol and/or drugs. It can be seen that for the most serious offences magistrates have a limited amount of discretion in levying a monetary penalty and determining the period of loss of licence, and that minimums are set by law. Convicted drivers may obtain a special (extraordinary) licence to enable driving under limited conditions (e.g. for hours of work) where a driver can show reason that such a licence should be issued (e.g. hardship such as loss of employment).

In addition to the above, a range of new offences and penalties are under consideration for specific classes of road users. For example, a national limit of 0.02 per cent for drivers of heavy vehicles and buses may soon be introduced. A lower limit may be also appropriate for commercial vehicle drivers who are transporting passengers, e.g. taxi drivers.

Alcohol and drug testing law

The Road Traffic Act allows for the testing of drivers for alcohol and other drugs. A police officer may request a driver to provide a sample of breath for analysis for alcohol where:

- the officer has reasonable belief that the driver is under the influence of alcohol; and
- random breath-testing (RBT) operations (introduced 1 October 1988) are underway.

Where a positive alcohol reading in excess of the legal limit results from such testing, the driver would be conveyed to the nearest major police station for additional testing using evidentiary breath-testing equipment. A reading in excess of the legal limit would provide sufficient evidence for the driver to be charged with a drink–driving offence.

Drivers stopped by police and suspected of driving under the influence of alcohol may be tested using passive or preliminary alcohol-testing devices. In Perth, use is made of Alcotech and Lion preliminary alcohol-testing devices. Some drivers who are stopped in random breath-testing operations may not be tested where impairment is not suspected, and allowed to continue driving. This situation is less likely today with the widespread use of quick and efficient breath-testing equipment used in the field.

Where a police officer suspects a driver to be impaired but a preliminary breath test reveals a negative or low alcohol reading, the officer may request the driver to provide a sample of blood for analysis for drugs. A sample is provided to police for analysis and for the driver who may wish to seek an independent analysis.

Laboratory testing for alcohol or drugs is conducted for the Police Department and the Coroner by the Chemistry Centre, WA Department of Minerals and Energy. This testing may be for samples from drivers fatally or seriously injured in a road crash, or for drivers suspected of driving while impaired. It is estimated that toxicology testing is performed on over 90 per cent of fatality cases in Western Australia. The testing includes the following drug classes: alcohol, amphetamines, other stimulants (ephedrines), benzodiazepines, cannabinoids, opiates, anti-depressants, anti-convulsants, and major tranquillisers. Testing is for all categories of drugs and is not discontinued when a positive alcohol level or evidence of one drug is recorded.

8.3 Current research

A Western Australian database

Road fatality data on drug involvement has been collected by the Chemistry Centre of Western Australia since 1990 and systems are being developed for linking this with other crash information held by the Police Department's Research and Statistics Unit. This will provide a basis for analysis and monitoring trends in fatal crashes.

A national database on drugs and driving

An Austroads Working Group is currently undertaking a study into the feasibility and methodology for establishing a national database of the evidence of drug use in drivers killed in road crashes. This project aims to identify existing data collections and sources, determine whether additional national data would significantly improve the state of knowledge, and investigate the feasibility of collecting this information in a regular and ongoing fashion.

Part of the above programme is likely to include an extension of responsibility analyses, undertaken first by the Victorian Institute of Forensic Pathology, to allow an assessment to be made of the driver's culpability or responsibility in an accident. Robertson and Drummer (1993) proposed that, if drugs present in a driver contributed to crash causation, it would be expected that they would be over-represented in culpable drivers. It was found that in 341 fatalities alcohol-positive drivers were statistically over-represented in the culpable group, in single-vehicle accidents and those accidents where vehicles left the road for no apparent reason. A larger sample size is necessary to assess the contribution of other drugs to crash involvement.

A pilot data collection project on responsibility analysis is soon to be conducted in Western Australia. This will involve the collection of information on drug use of drivers and motorcyclists involved in fatal crashes from 1990 to 1992. Data will be also collected on other factors involved in these crashes. This information together with that from other major Australian jurisdictions will assist in a better appreciation of the magnitude and contribution of drugs and driving in fatal crashes.

8.4 Further areas for attention

Simpson (1993) has identified a number of research and information needs relating to alcohol and drugs in transportation. Using this framework, it is suggested that further work be conducted on drugs and driving in Australia in the following areas:

- Data collected, e.g. non-fatal crashes: mandatory for all crash related hospital admissions and drivers involved in such collisions
- Data quality, e.g. standards and linking existing data sources
- Patterns of drug use, e.g. what is used, where, who by, and when
- Relationships with other factors, e.g. fatigue
- Effects on driving, e.g. safe operation of trucks
- Field testing devices, e.g. mass screening by police
- Impact of enforcement programmes
- Impact of education programmes

8.5 Conclusion

Currently, insufficient information is collected on the involvement of drugs and driving and road crashes. Limited information in Western Australia suggests that there may be increasingly more involvement of drug-affected drivers in serious road crashes, and that this area should be targeted for increased investigation in relation to data collection and analysis, programme development and enforcement, and evaluation of appropriate crash countermeasures.

Acknowledgments

Thanks are extended to the staff of the Research & Statistics Unit, Western Australian Police Department, and Chemistry Centre of Western Australia.

References

Campbell, N., 1990, *Cannabis and driving offences: Is there a case for a prescribed level?*, Perth: WA Chemistry Centre.
Homel, R., 1986, *Policing the drinking driver*, Canberra: Federal Office of Road Safety.
Homel, R., 1990, *High speed police pursuits in Perth*, Perth: WA Police Department.
Homel, R., Carseldine, D. and Kearns, I., 1988, Drink-driving countermeasures in Australia, *Alcohol, Drugs and Driving*, **4**, 113–44.
Lang, E. and Stockwell, T., 1991, Drinking locations of drink-drivers: A comparative analysis of accident and non-accident cases, *Accident Analysis and Prevention*, **23**, 573–84.
Lloyd, C., 1992, Alcohol and fatal road accidents: Estimates of risk in Australia 1983, *Accident Analysis and Prevention*, **24**, 339–48.
Robertson, M. and Drummer, O., 1993, Responsibility analysis: A methodology to study the effects of drugs in driving, *Accident Analysis and Prevention*, **26**, 243–7.
Simpson, H., 1993, Synopsis of discussion on research and information needs, *Alcohol, Drugs and Driving*, **9**, 1–5.
Stockwell, T., Maisey, G. and Smith, I., 1991, *Random breath testing in Western Australia – 2nd year review*, Perth: National Centre for Research into the Prevention of Drug Abuse, Curtin University of Technology.

Traffic Board of Western Australia, 1989, *Evaluation of the introduction of random breath testing (RBT) in Western Australia*, Perth: Traffic Board of Western Australia.

Traffic Board of Western Australia, 1993, *Random breath testing in Western Australia: 4th year of operation*, Perth: Traffic Board of Western Australia.

WA Police Department, 1993, *Road Crashes in Western Australia*, Perth: Traffic Board of Western Australia.

West, L. and Hore, T. (Eds), 1980, *An analysis of drink driving research*, Clayton, Victoria: Monash University.

9 Tracing the problem of driver fatigue

Peter Ingwersen*

9.1 Introduction

This chapter attempts to establish the extent of driver fatigue on Western Australia's roads using data extracted from WA Police Department accident report forms. The problem is felt to be largely underestimated due to a lack of reporting and/or a lack of candidness in reporting accidents involving driver fatigue. Other factors, such as limited interpretability of report form detail, and environmental factors, seem to conspire to misrepresent the extent of the problem. Essentially, analysis is restricted to single-vehicle accidents. Also, in order that some meaningful measure of the problem can be determined, consideration is progressively directed toward road sections felt to show an association with fatigue accidents, i.e. sealed, country and open roads.

9.2 Accident severity: Perth vs country Western Australia

In Western Australia, approximately 36 000 accidents were reported on average each year over the last five years, but of these, only 17 per cent occurred outside the Perth metropolitan region (Figure 9.1). Country accidents, however, account for 54 per cent of all fatal accidents and 38 per cent of all accidents involving a casualty admitted to hospital. Country accidents are more severe in nature than metropolitan accidents because of the higher vehicle speeds in country areas, and the slower response time of emergency service vehicles in attending crashes.

9.3 Under-reporting of accidents

The extent of the road accident problem however, goes beyond those which are reported to the police. First, it should be pointed out that, while all accidents involving a fatality or a casualty admitted to hospital are typically attended by the police, this is not the case for many less severe accidents. Recent work conducted by Rosman and Knuiman (1993) suggests that about 20 per cent of road accident casualties requiring hospital admission are not entered onto the Western Australian road accident database, and that perhaps as many as 50 per cent of property-damage-only accidents are not reported to the police. This work reveals that these under-reporting rates are substantially higher in country areas than for Perth, where it is typically only a five-minute drive to the nearest police station. It also indicates that accidents involving only a single vehicle are more likely to go unreported than those involving several vehicles.

Disclaimer: The opinions expressed in this chapter are those of the author and do not necessarily reflect those of Main Roads Department of Western Australia.

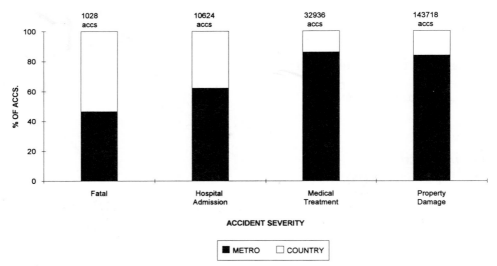

Figure 9.1   Metropolitan vs country accidents for 1988–1992.

9.4   Country accidents: open-road vs built-up areas

Using data extracted from Western Australia's road accident database for 1988–1992, the extent to which single-vehicle accidents are represented among all country open-road accidents is now considered. Nearly 60 per cent of country accidents reported the operative speed limit for the road section where the accident occurred. By confining the study to these accidents, and applying a threshold speed limit of 90 kph, these accidents can be assigned to either 'built-up' areas or 'open-road' areas. This facilitates investigation of the plight of single-vehicle accidents on the open road. Interestingly, using the 90 kph threshold, country accidents are approximately evenly split between built-up and open-road sections.

Figure 9.2 reveals that for both 'built-up' or 'open-road' roads, single-vehicle accidents are more strongly represented among casualty accidents than among property damage accidents. This fact indicates that multi-vehicle accidents, which typically occur at restricted speed limits of 60 or 70 kph, are less severe. It also reinforces the notion however, that single-vehicle accidents involving property-damage-only show a high level of under-reporting compared with multi-vehicle accidents. The most clear feature of this graph, however, is that nearly 80 per cent of 'open-road' casualty accidents involve a single vehicle only. Investigation also revealed that as many as 85 per cent of all country accidents known to involve driver fatigue occur on open-road sections.

9.5   Remote area accidents: WA vs south-west WA

To assess the impact of travelling on the open road in remote areas of the state, a breakdown of country single-vehicle accidents by region is provided (Figure 9.3). In the remote regions of Western Australia, the distribution of major roads and population centres is fairly sparse and the landscape less engaging compared with the south-west corner of the state. The south-west corner of Western Australia is loosely defined as that part of the state which is south and west of a line drawn between the towns of Kalbarri and Ravensthorpe. The lower vehicle densities of remote Western Australia, as one might expect, lead to a marginally higher prevalence of single-vehicle accidents

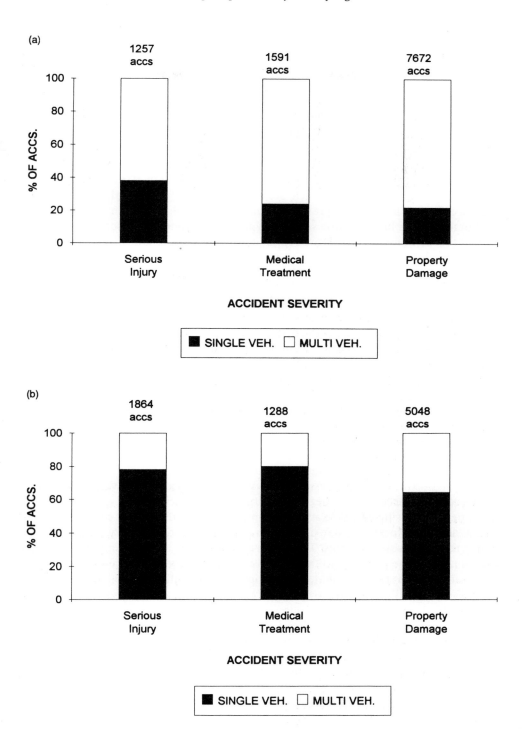

Figure 9.2 Country accidents in (a) built-up areas and (b) open-road areas, for 1988–1992.

in this region than for the south-west when considering all single-vehicle accidents. However, single-vehicle property-damage-only accidents show a relatively weaker presence in remote Western Australia than in the south-west. This is most probably due to lower reporting levels in the more remote regions of the state.

9.6 Accidents: sealed vs unsealed roads

The road surface material and road quality will, of course, affect the risk of having an accident. Figure 9.4 details accidents for single-vehicle accidents with respect to (a) sealed vs (b) unsealed open roads. It can be seen that on unsealed roads across all accident severities, single-vehicle accidents account for approximately 80 per cent of all accidents. For sealed roads, single-vehicle accidents overall account for about 65 per cent of all accidents. The stronger presence of multi-vehicle accidents on sealed roads is attributable to the higher traffic volumes typically encountered on sealed roads, and the fact that few unsealed roads are found in townsites, where multi-vehicle accidents are most likely to occur.

9.7 The hidden fatigue element in accidents

The characteristics of single-vehicle accidents have been provided as a preamble to the issue of identifying the extent of the driver-fatigue problem on country roads. Fatigue has been defined as 'a progressive decrement in performance which if not arrested, will end in sleep', and '... is related to the level of arousal ... of the driver' (Roads and Traffic Authority, 1990: 28). It is now appropriate to put in perspective driver-fatigue as a component of country accidents, compared with other contributing factors (Table 9.1).

It must first be stated however, that because of the similar nature of accident types, the Western Australian accident database includes within the driver fatigue category, accidents attributed to driver inattentiveness. Qualitative analysis suggests that inattention is relatively infrequent, and as accidents due to inattention cannot be distinguished they shall be classed as fatigue accidents.

Analysis of contributing factors to country open-road accidents reveals that driver-fatigue accidents account for 6 per cent of single-vehicle accidents on sealed roads and only 1 per cent of single-vehicle accidents on unsealed roads. Also, it can be seen that fatigue accidents form a higher proportion among the serious-injury accidents than among the less severe accidents. This suggests that fatigue accidents are typically of a serious nature, due probably to the fact that the driver does not take evasive action. However, it is also suspected that driver fatigue is more fully reported among serious injury accidents, albeit if partly by police conjecture, because these serious accidents are more likely to be attended by police.

The small number of fatigue-related accidents on gravel roads is not unexpected, as on these roads, drivers will typically be making short trips. Also, drivers on these roads will of necessity need to maintain an increased level of alertness to control even the more straightforward manoeuvres on gravel roads – this will reduce the potential for 'highway hypnosis'.

The figure of six per cent for accidents on sealed roads may however, understate the real incidence of fatigue-related accidents on sealed open roads. Detailed research conducted in 1987/88 by Armour *et al.* (1990) on serious-injury accidents on selected routes in Victoria suggested that in up to 33 per cent of single-vehicle accidents, fatigue was a contributing factor. Other research has suggested that fatigue may contribute to

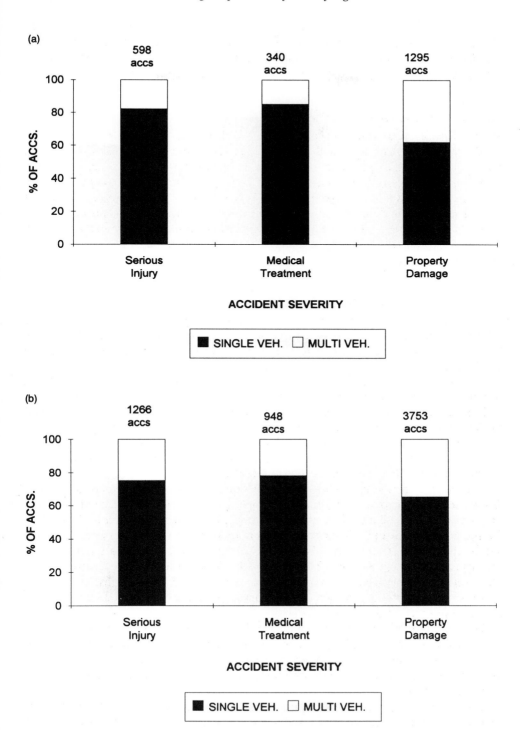

Figure 9.3 Accidents on open roads in (a) remote Western Australia and (b) the south-west of Western Australia, for 1988–1992.

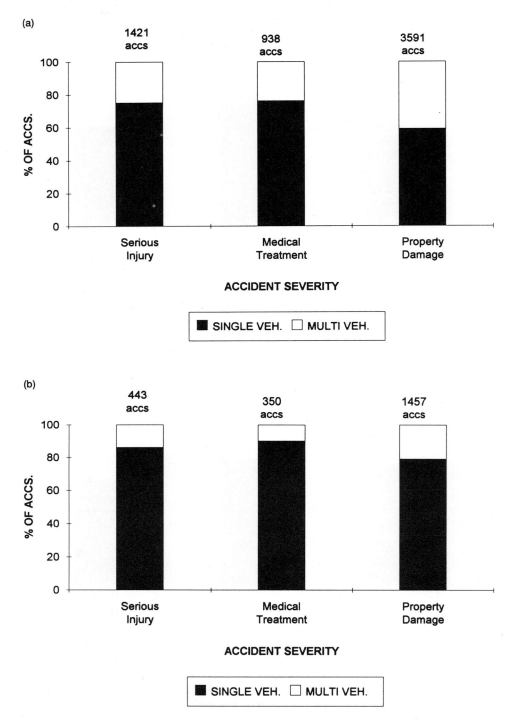

Figure 9.4 Country open-road accidents on (a) sealed roads and (b) unsealed roads, for 1988–1992.

Table 9.1 Country, open-road, single-vehicle accidents by movement type (1988–1992)

Movement type	Accident severity			
	Serious Injury %	Medical Treatment %	Property Damage %	All Severities %
Sealed roads				
Avoiding animal	8	11	16	13
Swinging wide at corner	11	7	5	7
Driver fatigue	**8**	**5**	**5**	**6**
Blowout	4	9	8	7
Lost control				
Road condition	1	1	1	1
Gravel shoulder	19	18	15	17
Loose gravel	1	2	2	2
Reason uncodeable	21	27	24	24
Reason unknown	18	8	7	10
Other	9	12	18	14
Total	100	100	100	100
No. of accidents	1067	716	2101	3885
Unsealed roads				
Avoiding animal	8	13	14	13
Swinging wide at corner	8	7	7	7
Driver fatigue	**2**	**1**	**1**	**1**
Blowout	3	5	5	5
Lost control				
Road condition	3	4	6	5
Gravel shoulder	4	4	6	5
Loose gravel	27	35	31	31
Reason uncodeable	20	17	16	17
Reason unknown	14	6	5	7
Other	10	8	10	9
Total	100	100	100	100
No. of accidents	379	314	1148	1841

between 35 per cent and 50 per cent of a
up to 70 per cent of heavy-vehicle accid

Now, while there is a large difference
and a factor 'causing' an accident, there
accidents are substantially under-report
base.

Firstly, it can be seen that 17 per cent o
open roads are classed as 'lost control c
ways, which have two lanes for 95 per cen
is 16 per cent. A further 9 per cent of cou
roads are recorded as 'lost control, reason
apply vehicle movement codes based upor
on the accident report form. In the abs

speculate as to what may have caused a vehicle to lose control. However, as a researcher, one must ask the question, 'If a vehicle lost control on the gravel shoulder, why was it there in the first place?' The data suggest that many less severe, fatigue-related accidents, are hidden among the more obscure accident causes such as 'lost control on gravel shoulder'. The accident reporter's vagueness may deliberately protect a driver's ego, or perhaps make a more mitigating testimony in support of anticipated vehicle insurance claims.

Second, close scrutiny of reported detail, in respect of fatal accidents, further reinforces the suspected problem of under-reporting of fatigue-related accidents. A study of country fatal accidents for a complete calendar year revealed 33 accidents for which, in the absence of any other discernible causes, fatigue is considered by the author to be a potential factor, although not recorded as such on the database. Fourteen of these accidents were on straight, level, dry, sealed sections of road. Seven of these 14 accidents were on well-formed highway sections, and a separate set of seven of these same 14 accidents describe the vehicle as veering to the wrong side of the road without any previous sign of driver over-correction. Four accidents sat in both camps, i.e. the vehicle veered to the wrong side for no apparent reason on a two-lane highway.

There is a lot of inference above, but it does make a strong case for the under-reporting among the more severe accidents of driver fatigue, or at least of driver inattention. Finally, anecdotal evidence from police officers suggests that fatigue plays a larger role in country accidents than is widely reported.

9.8 Comparison with other states

A comparison of accidents where vehicles have left the carriageway on country open roads is now presented for 1991 for four states: New South Wales, Victoria, Queensland and Western Australia. Accident data from the Federal Office of Road Safety's Serious Injury Database have been applied against vehicular travel data from the Australian Bureau of Statistics to provide accident rates per billion vehicle kilometres of travel for single-vehicle run-off-the-road type accidents (Figure 9.5).

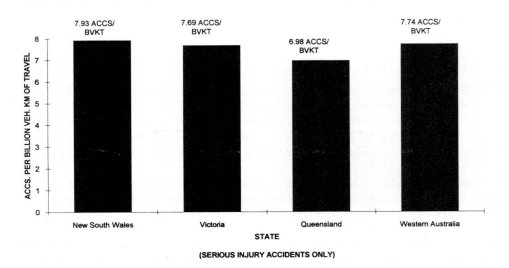

Figure 9.5 Accident rates by state for country, out-of-control accidents on open roads.

Table 9.2 Fatal accidents involving fatigue by state for 1988

State	Percentage of crashes involving fatigue
New South Wales	4.0
Victoria	5.0
Queensland	7.3
South Australia	4.1
Western Australia	**8.1**
Tasmania	4.8
Northern Territory	14.3
Australia Capital Territory	0.0

Source: Haworth and Rechnitzer, 1993.

Although one might have anticipated the larger and less populated states of Queensland and Western Australia to have shown higher rates for these accidents than New South Wales and Victoria, in fact there is little difference found between the states. Different application and interpretation of 'road user movement' codes by states may explain this unexpected result.

A comparison of the incidence of fatigue-related accidents for all states is presented in Table 9.2, courtesy of work done by Haworth and Rechnitzer (1993), who conducted a study for the Federal Office of Road Safety, of police reported detail supplemented by coronial inquest data for fatal accidents which occurred in 1988. The table lists for all Australian states and territories, the incidence of accidents which were attributable to fatigue as a percentage of all fatal accidents in each state. The Northern Territory, with 14 per cent, showed the highest proportion of fatigue accidents, followed by Western Australia with eight per cent and Queensland with seven per cent. These data reveal the strong association one expects to find between fatigue accidents and the states with typically longer distances between population centres, such as Western Australia and Queensland. However, notwithstanding the small numbers involved, it is interesting to note that Victoria and Tasmania both show higher proportions of fatigue accidents than New South Wales and South Australia.

9.9 Accident rates in Western Australia

Further consideration of the impact of distance travelled and its affect on accidents at a local level, is provided by re-visiting the analysis of south-west Western Australia vs the rest of the state. Open-road accident rates for fatigue-related accidents and single-vehicle accidents are provided for the state's highways and sealed main roads, where vehicle exposure rates are readily available (Table 9.3).

The accident rates (in accidents per billion vehicle kilometres of travel) shown across all severities, reveal that drivers in south-west Western Australia are 50 per cent more likely to suffer a fatigue-related accident than drivers in the remote parts of the state. This trend is strongly evidenced among the less severe accident categories. Also, fatigue accidents in the south west account for 6.6 per cent of all single-vehicle accidents compared with 4.7 per cent for remote Western Australia. These facts are surprising, even allowing for the more forgiving road environment and suspected lower reporting levels of accidents in remote Western Australia.

Table 9.3 Western Australia open-road accidents by severity for 1988–1992

Accident severity	Single-vehicle accident		Fatigue accident	
	Frequency	Rate	Frequency	Rate
Remote WA				
Serious injury	310	52.8	19	3.2
Medical treatment	172	29.3	4	0.7
Property damage	467	79.6	22	3.7
All severities	949	161.7	45	**7.7**
South-west WA				
Serious injury	481	43.1	44	3.9
Medical treatment	379	34	26	2.3
Property damage	1094	98.1	59	5.3
All severities	1954	175.2	129	**11.6**

Rates expressed as no. of accidents per billion vehicle kilometres of travel.

The trends may be a reflection of a higher prevalence of the inexperienced 'weekend tripper' encountered in the country areas nearer to metropolitan Perth. Drivers living in remote areas are more likely to be accustomed to long distance trips and the environmental conditions associated with remote driving, and hence better prepared to handle the problem of fatigue. This aspect is further discussed later in this chapter.

9.10 Geographical factors

New technologies such as geographical information systems (GIS) better allow for both the spatial analysis and integration of accidents with other related data. To investigate country road links showing a high propensity for driver fatigue accidents, it is useful to link location and road environment information.

Identification of road links most susceptible to driver fatigue is determined by considering both accident rates and accident frequencies for the road links. Frequency is included in the selection process to exclude those links showing an improbably high accident rate resulting from a chance 'few' accidents combined with a low exposure risk, that is, low traffic flow. A minimum of four fatigue accidents per 100 million vehicle kilometres of travel and three recorded fatigue accidents was adopted as a threshold value to provide meaningful results when identifying fatigue-accident prone links. No links meeting these minimum criteria were identified in remote Western Australia. Using these criteria, the most 'wearisome' road links on Western Australia's major road network appear to be located on highway routes and approximately 1–2 hours driving time from Perth. This lends some weight to the contention that city-based drivers are likely to be over-represented in fatigue-related accidents on country roads near to Perth.

The highways sections mentioned above, are all sealed to acceptable two-lane road width. While wide, sealed roads will be expected to reduce accident occurrence, this would not seem to be the case for fatigue accidents. High-quality pavement perhaps contributes to 'highway hypnosis', wherein high comfort levels combine with limited environmental distractions and long distances, to induce a drop in driver concentration.

By layering map detail of road links, reflecting a high propensity for fatigue accidents, onto a map detailing road seal widths, one can see that nearly all of the fatigue-accident prone links are on sealed two-lane highway sections.

GIS techniques can be utilized to assess the impact of these road links against other environmental factors including traffic volumes, road roughness, road alignment, and so on. Localized geographical features, such as density of townsites and roadside scenery, can also be analysed to investigate their impact on fatigue accidents. For example, the Perth Albany Highway which appears to be a particularly wearisome route, has moderate traffic volumes but offers little in the way of driver distraction, so that the road does not require high levels of concentration. Driving along this high-quality road from Perth, the driver is confronted with mostly similar countryside and no townships until reaching Kojonup, some 220 km south-east of Perth. The relative monotony of this trip and the moderately low traffic flows almost certainly contribute to the number of fatigue accidents. The Main Roads Department of Western Australia is pursuing these link analyses to determine placement of countermeasures to fatigue.

9.11 Day of week

Weekend drivers and their involvement in driver-fatigue accidents is now reconsidered.

Figure 9.6 reveals that most driver-fatigue accidents occur on Saturday and Sunday in both the south-west and remote parts of the state. The same is true of country single-vehicle accidents and all country accidents. However, if we look more closely at the

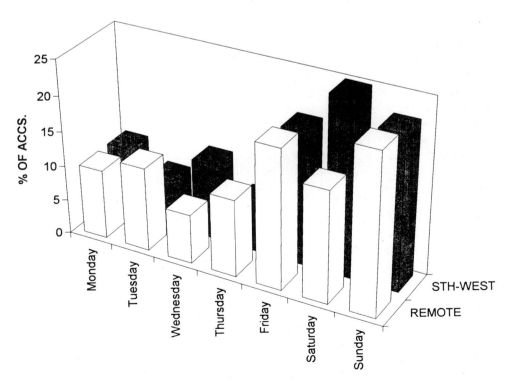

Figure 9.6   Fatigue accidents by region by day-of-week for 1988–1992.

Table 9.4 Country weekend accidents and traffic by region for 1988–1992

Accident type	Remote WA (%)	South-west WA (%)
Sat/Sun accidents as % all accidents		
All country accidents	30	32
Country single-vehicle accidents	37	39
Country fatigue accidents	**40**	**47**
Sat/Sun traffic as % all traffic	30	28

relative occurrence on Saturday and Sunday, of these three accident types (i.e. all accidents, single-vehicle accidents and fatigue accidents) and compare these with weekend traffic flows, two interesting patterns emerge (Table 9.4).

First, the relative proportion of accidents occurring at the weekend increases from a mean of 31 per cent for all country accidents to a figure of 38 per cent for country single-vehicle accidents to 45 per cent for country fatigue accidents. Clearly, weekend drivers present a higher risk than all drivers for both single-vehicle accidents and fatigue accidents, but more particularly for fatigue accidents. This is probably because these drivers have less experience of country driving than their weekday counterparts. The second pattern to emerge is that for fatigue accidents, there is a significantly higher presence of weekend drivers in the south-west of Western Australia than for remote Western Australia. This is possibly attributable to inexperienced, Perth-based, drivers going for a weekend trip to the country and succumbing to the demands of long-distance driving.

These patterns suggests that more work needs to be done to establish the origins of drivers involved in fatigue-related accidents. Recent road safety campaigns targeting driver fatigue have been presented exclusively on country television and radio networks. There is a possibility that these campaigns may be missing the mark ... that in neglecting the metropolitan audience, the major group of road accidents victims is being overlooked.

9.12 Time of day

Consideration of the time of occurrence of accidents reveals that over half of these, for all country Western Australia, occur between 7 p.m. and 7 a.m. when only about 20 per cent of vehicular travel is conducted (Table 9.5).

Although one associates weariness with the evening hours, the actual proportion of

Table 9.5 Country night-time accidents and traffic by region for 1988–1992

Accident type	Remote WA (%)	South-west WA (%)
Night accidents as % all accidents		
All country accidents	29	25
Country single-vehicle accidents	36	38
Country fatigue accidents	**58**	**51**
Night traffic as % all traffic	20	18

53 per cent of accidents is higher than one might expect, particularly when night-time accidents account for only 37% of country single-vehicle accidents. Fatigue accidents occur relatively more frequently at night in the remote regions than in the south-west. This is probably due to the longer distances that a night-driver must travel in order to reach a target destination in the more remote regions.

If the notion of the problematical 'weekend tripper' is now re-considered, it is suggested that fatigue accidents occurring between 4 p.m. on Friday and 7 p.m. on Sunday account for 52 per cent of all fatigue accidents in the south-west compared with 39 per cent (not shown) for remote WA. Nearly two-thirds of these 'weekend trippers' in the south-west are aged 25 years or under.

9.13 Driver age

Considering driver age (see Figure 9.7), it can be seen that that while the breakdown by age group is similar across both remote and south-west Western Australia, the 17–25 year old group present by far the highest risk in respect of fatigue accidents.

Of all single-vehicle fatigue-related accidents in country Western Australia, 51 per cent involved drivers of 17–25 years of age. Although this proportion is high, it is not greatly higher than the over-representation of young driver involvement in all accidents across all Western Australia, including Perth (Table 9.6).

A more interesting trend is that the presence of the younger driver is noticeably

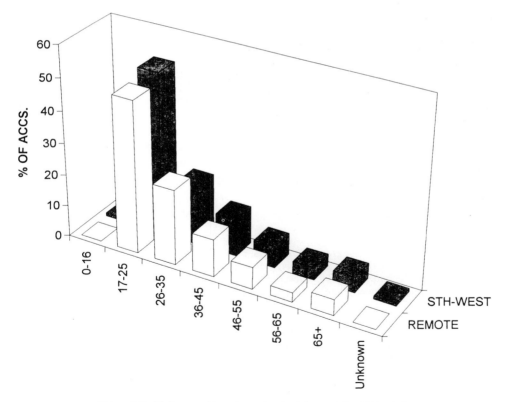

Figure 9.7 Fatigue accidents by region by driver-age for 1988–1992.

Table 9.6 Country accidents involving drivers aged less than 26 years

Accident type	Remote WA (%)	South-west WA (%)
All country accidents	45	46
Country single-vehicle accidents	43	48
Country fatigue accidents	**48**	**51**

1. Of all accidents (WA) 45 per cent involved drivers aged less than 26 years.
2. Of the total distance travelled (Australia), 13 per cent is by drivers aged less than 26 years.
3. Of licensed drivers (WA), 15 per cent are aged less than 25 years.

higher in respect of single-vehicle accidents (and the fatigue accident subset) in the south-west than in remote Western Australia. This reinforces the suggestion of a lower level of country driving experience among those driving in the south-west area of the state.

9.14 Countermeasures

A brief discussion of road environment countermeasures to fight driver fatigue is now provided. In Western Australia, trip distances are very long and stimulation from environmental settings and traffic volumes very low, so that drivers' concentration can easily lapse. Much of the major road network offers a relatively uninteresting landscape, such as infrequent townsites and landmarks to provide the driver with useful distraction. These factors are conducive to driver fatigue. The Main Roads Department of Western Australia is developing techniques, some under trial elsewhere, to promote a staged and cost-effective solution to the impact of fatigue on driving.

This chapter has identified that most fatigue occurs on high-quality routes in the south-west of the state, such as Perth-Albany Highway. These well-utilized but tedious, routes may need to convey safety messages on road signs or billboards, which should be complemented by strategically placed and well-appointed roadside rest areas. Efficient use of profile edge-lining and rumble pads should be made on particularly wearisome road sections. Campaigns, encompassing both the media and community groups, should be encouraged to alert drivers, particularly high-risk drivers, to this subtle but serious problem of driver fatigue.

References

Armour, M., Carte, R., Cinquegrana, C., Francis, P. and Griffith, J., 1990, Study of single vehicle rural accidents, *Driver Factor's Report*, Melbourne: VICROADS.
Haworth, N. and Rechnitzer, G., 1993, Description of fatal crashes involving various causal factors, *Driver Factor's Report*, Canberra: Federal Office of Road Safety.
Hulbert S., 1972, Driver fatigue, in Forbes T.W. (Ed.), *Human Factors in Highway Traffic Safety Research*, Ch. 12, New York: John Wiley.
Roads and Traffic Authority, Road Safety Bureau, New South Wales, 1990, Report of Proceedings on 'Workshop on Driver Fatigue'.
Rosman, D.L. and Knuiman, M.W., 1993, A comparison of hospital and police road injury data, Road Accident Prevention Research Unit of Western Australia.
Shuman, M., 1992, Asleep at the wheel, *Traffic Safety*, Jan/Feb, 1992.

10 The interaction between driver impairment and road design in the causation of road crashes – three case studies

Patrick J. Kenny

10.1 Introduction

In any road crash, the cause may be attributed to a combination of defects or defective performance in a number of factors. These factors are usually grouped under three headings (Grime, 1987):

- Road environment
- Road user
- Vehicle

The road environment covers the area of geometric road design, as well as the elements of road furniture such as: signs, road markings, crash-rail and traffic signals. British research has shown this factor to be present in 28 per cent of road accidents, although it was solely responsible for only 2.5 per cent of accidents (Sabey and Straughton, 1975).

The same study found that defective vehicle performance was found to be present in 9 per cent of road accidents, although it was the sole cause of only 2.5 per cent of road accidents.

Human factors were found to be present in 95 per cent of all road accidents. In 65 per cent of road accidents, human factors were judged to be the sole cause of the accident. Since the driver can vary the difficulty of the task by adopting a slower speed or by selecting lower risk options, it is not surprising that, in hindsight, it can be seen that the driver may have avoided the crash. Thus the accident is attributed to driver error.

Human factors that have particular relevance to road accident analysis include:

- driving while under the influence of alcohol;
- failure to wear seat-belts;
- selection of incorrect driving speed, and
- inability to perceive visual information, either through poor eye-sight or inattention.

The failure to make correct decisions regarding speed or to perceive important visual cues may be due to impairment of the driver's faculties due to fatigue.

The concept of fatigue, while acknowledged as a potentially important element in the cause of road accidents, does not have an agreed definition. Haworth et al. (1988) state that: 'Although the term "fatigue" does not have an accepted definition, fatigue has subjective, objective (performance) and physiological components which may occur in the short term or as a chronic state.'

Bartley and Chute (1947) made the distinction between 'impairment' which can be measured in physiological terms, and the term 'fatigue', which is used to describe the psychological aspects of the phenomenon. Brown (1993) defined psychological fatigue as a subjectively experienced disinclination to continue performing the task at hand, which has adverse effects on human efficiency largely when individuals continue working after they become aware of it. It is this definition which will be used here in the investigation of the interaction of fatigue and road design.

While road accidents are primarily a consequence of human error, the designers of the road network must take care to produce road systems (both roads and roadside environments) which can tolerate less than optimal behaviour by the road user. As Lay (1990) states: 'The design professions have an obligation to design for human error and not to piously condemn the next round of accident victims.' This chapter examines three road accidents which have occurred in New South Wales:

- a loss of control accident involving a small sedan colliding with a power pole in a rural village;

- a loss of control accident involving a large sedan colliding with a power pole in a regional city; and

- an accident involving a small sedan striking the third-last carriage of a moving train at night.

This chapter examines the road design and the fatigue/impairment aspects of these three accidents and makes recommendations regarding existing road design guidelines to cater for the less than optimal performance of the driver.

10.2 Case one – loss of control accident in a rural village

Brief description
The driver lost control of the motor car on a curve to the right. The car travelled to the left of the road and struck a power pole. No other vehicle was involved.

Road environment details

Weather:	fine
Time:	8 p.m.
Light:	street lights at approximately 50-m intervals
Environment:	rural village, residential area
Speed zone:	60 kph
Curve direction and radius:	curved to the right, 25 m radius
Speed prior to striking pole:	45–55 kph (estimated from damage to vehicle)
Curve approach speed:	60 kph
Speed zone 500 m before curve:	100 kph
Number of lanes:	2
Pavement width:	7.1 m
Super-elevation:	8 per cent
Distance from edge of road to light/power pole:	1.2 m
Pavement condition:	bitumen seal with some loose gravel present
Permanent warning signs:	none
Site accident history:	five years prior to this accident, one loss of control accident and one collision involving two vehicles

Driver details

Age:	21 years
Sex:	female
Driving experience:	5 years
Blood–alcohol level:	nil
Driving time prior to accident:	3 hours

Car details

Model:	1976 Gemini sedan
Condition:	no obvious defects.

Injury details
A 25-year-old male (passenger in the rear seat), suffered severe head injuries resulting in paralysis of the left leg and hand, speech defect and intellectual brain damage.

Police action
The police charged the driver with negligent driving.

Discussion
This accident occurred on a fine night, on a reasonably well lit street. There were no other motor vehicles involved. The vehicle was not speeding and the driver was not under the influence of alcohol. The driver had five years driving experience. The vehicle's tyres, brakes and steering do not appear to have contributed to the accident.

The factors which may have caused the accident are as follows:

- The design speed of the curve
  Using the design method of the Road Design Guide of the Roads and Traffic Authority, New South Wales (Roads and Traffic Authority, 1988), the design speed of the curve is 35 kph. This is significantly lower than the estimated approach speed of the accident vehicle.

- The failure of the local road authority to place a curve warning sign and advisory speed sign in advance of the curve

- Resealing of the pavement which occurred on the day of the accident
  This would have resulted in the road surface having some loose gravel on the surface. This would have increased the probability of losing control on the curve. However, no other accidents occurred on this curve after the resealing.

- The location of the power/light pole only 1.2 m from the edge of the pavement
  This would have increased the likelihood of a collision. Once having lost control, the driver had very little time to recover before hitting the pole (approximately 0.5 seconds).

- The fatigue of the driver
  Travelling for three hours prior to the accident and travelling in the evening, increased the likelihood of an accident. The fatigue resulted in her being ill-prepared for a tight curve occurring soon after leaving a 100 kph road. The arousal theory

of fatigue stems from research which shows that performance follows an inverted U-shaped curve. Performance is poor when arousal is weak or very strong and peaks when arousal is at a moderate level. Haworth *et al.* (1988) suggested that fatigue or boredom resulting from under-arousal is likely to result from continued rural or highway driving. In this case, the under-arousal resulting from the rural road journey just prior to the accident, probably led to a loss of attention when confronted with a road task of some complexity (manoeuvring a car around a tight curve at night, on a street with a slippery surface and with obstacles only 1.2 m from the pavement edge).

10.3 Case two – loss of control accident on a street in a regional city

Brief description
The driver lost control of a motor car on a curve to the right. The car travelled across the road and struck a power pole. No other vehicle was involved.

Road environment details

Weather:	fine, at time of accident – 4.6 mm of rain had fallen in the six hours prior to the accident
Time:	5.35 p.m.
Light:	daylight, street lights not operating
Environment:	regional city, light industry and storage yards
Speed zone:	60 kph
Curve direction and radius:	curved to the right, 100 m radius
Speed prior to striking pole:	60 kph (driver's estimate)
Approach speed to curve:	80 kph
Speed 500 m before curve:	80 kph
Number of lanes:	2
Pavement width:	13 m
Super-elevation:	1.7 per cent
Distance from edge of road to light/power pole:	0.2 m
Pavement condition:	asphaltic concrete, good condition
Permanent warning signs:	right curve sign
Site accident history:	in the five years prior to this accident, three loss-of-control accidents, all on wet roads

Driver details

Age:	35 years
Sex:	female
Driving experience:	7 years
Blood–alcohol level:	nil
Driving time prior to accident:	5 minutes
Work duration prior to commencing driving:	10.5 hours (including breaks for lunch and short breaks in morning and afternoon)

Car details
Model: 1983 Ford Falcon sedan
Condition: no obvious defects

Injury details
The driver suffered fractured ribs and internal chest injuries.

Police action
The police charged the driver with negligent driving.

Discussion
This accident occurred on a wet road, in daylight. There were no other motor vehicles involved. The vehicle was not speeding and the driver was not under the influence of alcohol. The driver had seven years driving experience. The vehicle's tyres, brakes and steering do not appear to have contributed to the accident.
   The factors which may have caused the accident are:

- The design speed of the curve
  Using the design method of the Road Design Guide of the Roads and Traffic Authority, New South Wales (Roads and Traffic Authority, 1988), the design speed of the curve is 55–60 kph. This also is significantly lower than the estimated approach speed of the accident vehicle (80 kph). Because the street is located in an urban area, the minimum desirable design speed of the curve is 60 kph. However, in this case, the road preceding the curve is a long (approximately 1 km) straight, on a downhill grade, with no cross streets. Even though the curve is appropriate for a design speed of 60 kph, the approach actual speed of most motorists is estimated to be of the order of 80 kph. The RTA Guide recommends that the design speed of curves should not be 10 kph less than the approach speed. Therefore, in this case the curve is under designed for the actual approach speed. The wet condition of the road would have increased the probability of losing control on the curve. This is reinforced by the accident history which shows three similar loss of control accidents in wet weather.

- The location of the power/light pole, only 0.2 m from the edge of the pavement
  This would have increased the likelihood of a collision. Once having lost control, the driver had very little time to recover before hitting the pole.

- Driver fatigue
  The driver had been working in a factory for ten hours prior to the accident and was then travelling in the late afternoon, fatigued. This resulted in her being unable to negotiate the curve successfully, which occurred at the end of a long straight stretch. Although the period of driving was actually quite short, the time on duty exceeded 10.5 hours. Most studies on fatigue and driving have examined the relationship between the time on duty of truck and bus drivers and their fatigue levels. Harris and Mackie (1972) found an increasing ratio of obtained to expected crashes between the seventh and tenth hours of driving. If work in a factory may be considered to contribute to fatigue as much as driving, then the driver in this instance could be expected to be at risk from a fatigue-induced accident. The Road Safety Bureau

of New South Wales (Road Safety Bureau, 1992) found in a survey of drivers that 20 per cent of fatigue-related accidents occurred less than half an hour after the trip commenced. Thus fatigue could be a factor in this accident, even though it occurred at the early part of the journey.

10.4 Case three – collision with the end of a moving train in a rural town

Brief description
The driver struck the third-last carriage of a moving train. He appears to have misjudged the location of the end carriage.

Road environment details

Weather:	fine
Time:	11.10 p.m.
Light:	street lights operating
Environment:	rural town, light industry and parklands
Speed zone:	60 kph
Curve direction and radius:	curved to the right, 70 m radius
Speed prior to colliding with train:	20 kph (driver's estimate)
Curve approach speed:	55 kph
Speed 500 m before curve:	75 kph
Number of lanes:	2
Pavement width:	9 m
Super-elevation:	0 per cent
Pavement condition:	bitumen seal, smooth condition
Permanent warning signs:	standard rail crossing signs
Rail level crossing control:	flashing lights and bells (checks indicate that they were in working order at the time of the accident)
Site accident history:	in the five years prior to this accident, none reported

Driver details

Age:	17 years
Sex:	male
Driving experience:	5 months
Blood–alcohol level:	negative report, however alcohol was consumed on the night of the accident
Driving time prior to accident:	5 minutes
Activity prior to accident:	dinner with group at restaurant starting at 7.30 p.m.

Car details

Model:	1978 Holden Gemini sedan
Condition:	vehicle seriously affected by corrosion (police report)

Injury details
Two passengers suffered severe head and spinal injuries.

Police action
The police charged the driver with culpable driving.

Discussion
This accident involved a car colliding with a moving train, which the driver knew was on the line. There were no other motor vehicles involved. The vehicle was not speeding. The driver claims that he did not see the last three unloaded carriages of the freight train. He also claims that the warning lights at the crossing had ceased flashing when he attempted to cross the rail line. Tests by the State Rail Authority have shown that this is highly unlikely. The driver had five months driving experience and had consumed some alcohol (at least 500 ml of beer). The vehicle was seriously affected by corrosion, but this would not have contributed to the occurrence of the accident.

The design of the road and the level crossing control does not appear to have been a factor in this accident since the driver was aware of the train's presence on the track. The driver's misjudgement may be explained as a withdrawal of attention from the road and driving tasks which Brown (1993) describes as the main effect of fatigue.

The factors which may have caused the accident are:

- misjudgement resulting from fatigue
  This type of accident is difficult to explain other than in terms of driver misjudgement resulting from fatigue. Even his lack of driving experience cannot explain his decision to drive into a moving train which he had seen. The accident occurred at 11.10 p.m. when fatigue accidents are more likely to occur. In addition, some alcohol was consumed by the driver. The driver is described as of 'slight build' and a small amount of alcohol may have contributed significantly to his apparent withdrawal of attention from the driving task.

10.5 Conclusions

In this chapter, three road accidents involving motor cars have been studied. Each driver made an error of judgement either in negotiating a curve or crossing a railway line. At first sight, the accidents are inexplicable, since few other accidents occurred at these sites and no other vehicle was involved.

Fatigue is often regarded as a problem for the professional driver. However, a recent study (Road Safety Bureau, 1992) has shown that one third of non-professional drivers have experienced an accident, or near accident, due to driver fatigue. In the accidents examined here, fatigue may have resulted from:

- lack of stimulation prior to reaching the accident site;
- long duration of work prior to commencing the trip; and
- withdrawal of attention due to driving late at night and alcohol consumption.

Road design guidelines for the geometric design of curves attempt to provide for less than optimal behaviour of the driver by recommending minimum values for curve radius, transverse friction and super-elevation. However, in order to mitigate the effect of fatigue on driver behaviour, road designers should:

- consider the effects of fatigue in the course of ordinary driving tasks, rather than assuming that it is only related to rural roads, long-distance travel and professional or touring drivers;
- adequately signpost sub-standard curves, particularly in areas close to high speed roads (e.g. small rural towns and villages);
- provide a wide 'clear zone' adjacent to the road so that the fatigued driver has time to correct an error before colliding with an object; and
- design curves which assume the actual approach speeds adopted by drivers, rather than the speed which conforms to the legal speed limit.

References

Bartley, S.H. and Chute, E., 1947, *Fatigue and Impairment in Man*, New York: McGraw Hill.

Brown, I.D., 1993, Driver Fatigue, Preprint of paper to be published in *Human Factors*.

Grime, G., 1987, *Handbook of Road Safety Research*, London: Butterworths.

Harris, W.H. and Mackie, R.R., 1972, *A study of the relationships among hours of service and safety of operations of truck and bus drivers*, (Report No. BMCS-RD-71-2), Washington D.C.: Bureau of Motor Carrier Safety (NTIS No. PB 213–963).

Haworth, N.L., Triggs, T.J. and Grey, E.M., 1988, *Driver Fatigue: Concepts, Measurement and Crash Countermeasures* (CR72), Canberra, Australia: Federal Office Road Safety.

Lay, M.G., 1990, *Handbook of Road Technology*, 2nd Edn: 607, New York: Gordon and Breach.

Road Safety Bureau, 1992, *Driver Fatigue – A Survey of Northern Region Drivers' Attitudes and Reported Accident Behaviour*. Road Safety Bureau Consultant Report CR9/92, for the Roads and Traffic Authority of New South Wales.

Roads and Traffic Authority, New South Wales, 1988, *Draft Road Design Guide*, Sydney: Roads and Traffic Authority.

Sabey, B.E. and Straughton, G.C., 1975, Interacting roles of road environment, vehicle and road user in accidents, *Fifth International Conference of the International Association for Accident and Traffic Medicine, and the Third International Conference on Drug Abuse of the International Council on Alcohol and Addiction*, London: September.

SECTION THREE:
Countermeasures to the adverse effects of fatigue on driving

11 The road to fatigue: circumstances leading to fatigue accidents

Dallas Fell

11.1 Introduction

Context

Road safety authorities in New South Wales have investigated the issue of driver fatigue since at least the early 1970s, with work addressing both theoretical analysis (Cameron, 1973) and practical effects, e.g. from results of in-depth accident investigations involving heavy vehicles (Linklater, 1978).

The NSW approach to the fatigue problem was no doubt shaped by the early orientation towards heavy vehicles, with their long hours of driving. Long hours are a major risk factor for driver fatigue and have been a focus of countermeasure messages through to the present time.

As fatigue emerged as a substantial risk factor for road accidents, community action and mass media advertising began. In NSW, community groups established roadside 'coffee stops' for holiday drivers in the early 1980s. The first driver fatigue advertisements appeared on NSW television in 1986.

Since that time, fatigue has become a major road safety campaign topic, along with drink–driving, occupant restraint use and speeding.

Sources of data on driver fatigue

Driver fatigue in road accidents is notoriously difficult to detect or measure – or indeed, to define! This uncertainty follows through to data on the incidence of accidents involving fatigue.

In some cases in NSW, police officers attending accidents will record that a driver was 'asleep', 'drowsy' or 'fatigued'. These comments generally arise from admissions of drivers or passengers or witness reports. Few accidents are recorded in this way each year, so they obviously do not indicate the size of the fatigue-accident problem, but they can be used to explore the characteristics of fatigue accidents, answering questions about drivers, locations and time of the accident.

To gauge the size of the fatigue-accident problem, the Road Safety Bureau has developed criteria for identifying accidents which may have involved fatigue. These include those identified by the police, plus accidents in which the vehicle left the road or crossed to the wrong side of the road for no apparent reason.

A further source of information on fatigue-accident prevalence and characteristics are in-depth accident studies. These can provide more information than usual accident reports on the state of the driver, but are still limited by the difficulties of detecting the presence of fatigue following an accident.

Gaps in the data

From routine police accident reports and in-depth accident studies, information is available on many features of accidents thought to involve fatigue. For instance, the 1992 accident data shows that 43 per cent of fatigue accidents occur in rural areas compared to 13 per cent of non-fatigue accidents.

Characteristics of fatigued drivers involved in accidents can also be explored. About half of the fatigued drivers in accidents are aged 25 years or under. They are usually driving a car (85 per cent) and usually alone in the vehicle (71 per cent). Most of these drivers are male (84 per cent).

Data on the characteristics of the accidents themselves show that they are more prevalent on weekends (48 per cent), they are more likely than other accidents to occur at night (54 per cent between 6 p.m. and 6 a.m.), they tend to involve one vehicle only (79 per cent) and that they are more likely than non-fatigue accidents to result in a death or a serious injury (20 per cent as compared to 11 per cent for other accidents).

This information gives us a sketch of a fatigue accident but there are many questions left unanswered. Where were the drivers going when they had their fatigue accidents? Where were they coming from? What time of day did they start out? How long had they been driving when the accident occurred? Did they take rest breaks? Had they had enough sleep?

11.2 The fatigue accident telephone survey

Survey method

We decided to try to fill some of the gaps in fatigue data by identifying a group of drivers who had been involved in fatigue accidents, through a telephone survey, and to investigate the circumstances of their accidents. We expanded the group to include people who had had a near-accident experience due to fatigue. In many cases, these situations would closely resemble those in which an actual accident occurred, and would enlarge our study group.

The study was carried out in the RTA's Northern Region of New South Wales. This area has high numbers of driver-fatigue accidents and regional staff were keen to explore the fatigue problem in the area. The region covers the area north of Sydney to the Queensland border and includes the Pacific and the New England Highways. The study and questionnaire, shown at the end of this chapter, were developed jointly with the NRMA. The survey and data analysis were contracted to Yann Campbell Hoare Wheeler, whose report on the survey has been published by the Road Safety Bureau (1992).

We surveyed 1000 drivers. Quotas were set for sex and age groups to ensure that there were adequate numbers for comparison. It should be noted that this means the sample is *not* a representative sample of NSW Northern Region drivers.

Drivers were asked about their knowledge of fatigue countermeasures, recall of advertising and advertising messages and their experience of driver fatigue accidents or near-accidents.

Characteristics of the most recent fatigue accident or near-accident trip were explored. Drivers who claimed they had not had a fatigue accident or near-accident were asked similar questions about their most recent trip of two hours duration or longer for comparison with accident trips. Differences between 'safe' trips and accident or near-accident trips were to be examined to help to identify factors which may be associated with a risk of fatigue accidents.

11.3 Survey results

Accident experience

Of the 1000 drivers in the sample 28 per cent (280) had had an accident or near-accident which they considered to be caused by their own driver fatigue, with 4 per cent having had an actual accident and 24 per cent having had a near-accident. Although accidents, as such, tended to be 'one-off', half of those who had had a near-accident had had more than one.

Although the sample of drivers was split evenly over the age groups, drivers who claimed to have had fatigue accidents/near-accidents tended to be those between 21 and 39 years of age. It would be expected that older drivers, given their longer exposure to driving, would be more likely to have had accidents. It might be assumed that older drivers are failing to report previous fatigue incidents (Table 11.1).

Trip purpose

Although holidays was one of the main reason for trips in which fatigue accidents/near-accidents occurred, other reasons were also prominent, such as family, social and even business.

Home was the most common starting place for all trips, but less often for accident/near-accident trips, which were more likely than other journeys to start from work, hotels/motels and tourist attractions. Apart from home, the most common destinations for accident/near-accident trips were friends' or relatives' house and usual workplace. Hence, the most common accident/near-accident trips were from home to friends' or relatives', friends' or relatives' to home and work to home. A point of interest here is that non-accident drivers are more likely to be heading away from home, whereas it is more common for the drivers in an accident/near-accident journey to be headed home, after social activity or work and so forth.

Who, where and when?

The fatigue accidents/near-accidents tended to involve drivers who were alone in the vehicle, travelling on highways or main roads, in rural areas. Starting times for fatigue

Table 11.1 Profile of the sample of drivers surveyed

Age in years	Total percentage $n = 1002$ %	Had accident $n = 37$ %	Had near accident $n = 243$ %	No accident $n = 642$ %
17–20	13	3	10	12
21–25	15	19	17	15
26–29	14	27	20	13
30–39	14	14	22	12
40–49	14	11	12	16
50–59	14	13	10	16
60–69	14	13	9	16
Mean age	35.2	38.4	36.1	40
Gender				
Male	50	81	72	46
Female	50	19	28	54

accident/near-accident trips were spread across the day, but the proportion starting late at night or early in the morning was high compared to non-accident trips. Accident times were similarly spread, with almost half of the accidents/near-accidents occurring between 6 p.m. and 6 a.m.

Fatigue prevention measures and risk behaviour

In line with the high proportion of solo drivers, drivers on accident/near-accident trips were less likely to share the driving, with only 13 per cent sharing the driving, compared to 30 per cent of drivers on other trips. The patterns of rest by drivers on accident/near-accident trips were different from those on non-accident trips, with accident/near-accident drivers less likely to take breaks, and when they did take breaks, taking fewer and, on average, shorter breaks. Main reasons given for not stopping were 'wanted to get the journey over with' and 'did not want to waste time'.

As would be expected, the accident/near-accident drivers were more likely to have experienced driver fatigue. However, while little more than half the drivers on accident/near-accident trips acted to avert fatigue, 61 per cent of the drivers on an average trip of two hours or more took action to avert fatigue, though only 29 per cent actually experienced fatigue. The action taken differed markedly between the two groups, also, with accident/near-accident drivers favouring winding down the window and playing the radio to avert fatigue, and drivers on ordinary trips relying on taking breaks.

More of the accident/near-accident drivers had consumed alcohol before the trip. The drivers on accident/near-accident trips were also far more likely than drivers on ordinary trips to have missed out on a good night's sleep before the journey.

Driving time

Fatigue is often spoken of in association with long distance driving, but 59 per cent of the accidents and near-accidents in this study occurred within two hours of the start of the trip. The shortest fatigue accident/near-accident trip was only three minutes long. Factors other than long hours of driving are obviously influential in the development of driver fatigue. On average, drivers had spent between two and three hours at the wheel.

Knowledge of driver fatigue and awareness of advertising

Lack of attention and difficulty concentrating was the most popular description of driver fatigue, rather than descriptions such as falling asleep. This shows a tendency to associate driver fatigue with early indications, rather than extremes like falling asleep at the wheel. This was true for both accident and non-accident drivers.

Agreement was almost universal that driver fatigue accidents can be avoided, and key avoidance tactics for both accident and non-accident drivers were 'stop driving when tired' and 'take regular breaks'. Accident-involved drivers seemed less aware of the importance of breaks from driving, not driving for too long and avoiding driving at normal sleeping times (Table 11.2).

Over 90 per cent of respondents could recall driver fatigue advertising, in both

Table 11.2 Driver's knowledge of things to do to avoid fatigue-related accidents

	Total percentage n = 972 %	Had Accident n = 36 %	Had near Accident n = 235 %
Do not drink when you are tired	95	92	94
Have a good night's sleep the night before driving	95	89	94
Stop driving when you feel tired; rest, walk around	94	92	91
Take regular breaks from driving, rest and walk around	94	83	94
Share the driving	94	89	95
Do not drink alcohol	92	89	91
Do not drive for too long	92	83	91
Plan your trip	88	89	84
Wind down the window	62	61	62
Do not drive at times when you would normally be asleep – late in the evening or early morning	57	39	52
Eat during the trip	55	50	59
Drink tea, coffee or Coke	45	42	46
Play and turn up the volume of your radio or tape	30	19	32
Other	3	3	3

accident and non-accident groups. Key advertising messages recalled were:

- Take regular breaks.
- Stop driving when you feel tired.
- Stop, revive, survive.
- Don't drive when tired.

11.4 Conclusion and discussion of survey

Methodological issues

The comparison group of drivers on trips of two hours or more was chosen on the assumption that most fatigue-accident trips would tend to be long-distance trips. The fact that many accidents or near-accidents happened within the first two hours casts doubt on the usefulness of this group for comparison, especially as the intended length of the trips for accident/near-accident drivers is unknown. Issues such as the taking of breaks or the sharing of driving need further exploration to confirm that differences between the groups would hold for comparable trip lengths.

General

A positive outcome of the study was that attitudes and knowledge of respondents were closely in line with most of the driver-fatigue messages which are propounded by the RTA, the NRMA and others. The concern which emerges is that, although drivers involved in fatigue accidents/near-accidents had a similar level of knowledge about fatigue as other drivers, the accident trips often involved behaviour which was not in keeping with this knowledge.

Returning to our original questions to fill gaps in available data, we have managed to fill some of these gaps with new knowledge and for some, we have new questions.

Bearing in mind the methodological limitations mentioned above, answers were as follows.

Where were the drivers going when they had their fatigue accidents or near-accidents?
About a quarter were going to or from holidays, and others on family, social or business trips.

What time of day did they start out?
Journey starts were spread throughout the day, but a third of the accident/near-accident journeys started between 8 p.m. and 6 a.m. compared to 19 per cent of the comparison journeys.

How long had they been driving when the accident occurred?
Of the drivers 20 per cent had driven for longer than five hours but many fatigue incidents were occurring in the first two hours of the trip.

Did they take rest breaks?
Generally drivers had few and short rest breaks, but more than half had not travelled for two hours, clouding the meaning of the lack of breaks.

Had they had enough sleep?
For almost half, the answer was no.

Often involving solo drivers, fatigue precautions before and during the trip were scanty. Wanting to finish their trip, accident/near-accident involved drivers wound down the window or turned up the radio rather than stopping for rest. Although further examination of short accident/near-accident journeys is needed, the research seems to support the importance of the following factors in avoiding fatigue accidents:

- sharing the driving;
- taking breaks;
- taking rest breaks, taking more breaks and longer breaks;
- the preventative role of rest breaks;
- having a good night's sleep before driving; and
- not driving at night (during normal sleeping times).

The research raised the following new issues:

- Other fatigue risk factors can cause accidents before long hours of driving take effect (e.g. inadequate sleep).
- Fatigue is not just (or even primarily) a holiday-driving problem.
- Drivers who have had fatigue accidents or near accidents know the fatigue messages

and express agreement with them but did not act in accordance with them on the trip in question.

11.5 New directions in driver fatigue: post-research

Further research

As is often the case, this research has raised as many questions as it has answered. Far from showing the classic long hours of holiday-driving pattern, the fatigue-accident target group has fragmented into a variety of groups. More information can be gleaned about these groups from the survey data, and further research will be needed to investigate fatigued drivers on non-holiday trips.

Another issue of interest for further research will be the factors causing drivers to act against their knowledge and beliefs about fatigue avoidance on particular trips, and whether these breaches were 'one-off' or regular behaviour for accident/near-accident drivers.

Better selection of comparison groups would be warranted for future research in this area.

Changes to countermeasures

From the research, it appears that new messages and target groups will be needed for fatigue messages. For instance, messages about risk factors such as inadequate sleep are needed to supplement and balance messages on long hours of driving. New target groups will need to include drivers on social and work trips, once more is known about these groups.

11.6 Summary of questions asked of drivers

Demographics

- Sex
- Age

Knowledge of driver fatigue countermeasures

- Which *one* of the following do you think best describes driver fatigue: falling asleep at the wheel, tiredness, drowsiness, nodding off, lack of attention/difficulty concentrating, other (specify)?
- Do you think there is anything you can do to avoid fatigue accidents?
- (If yes) What sorts of things do you think you can do to avoid fatigue accidents?
- Now I am going to read you a list of things other drivers have mentioned to avoid fatigue accidents ... which ones do you think you could use to avoid fatigue accidents?
- Have you ever seen or heard any advertising about driver fatigue?

- What do you think are the main things driver fatigue advertising is trying to tell drivers? (Probe fully.)

Specific questions about your experiences with driver fatigue
- Have you ever had an accident because of tiredness or fatigue? How many?
- Have you ever had a near-accident because of tiredness or fatigue? How many?
- Have you ever driven on a journey that took more than two hours of driving?

Questions about the most recent accident/near-accident journey or journey of two hours or more

Questions marked with an asterisk (*) were reworded appropriately for the drivers describing their trip of two hours or more.
- Where did the journey that day originally start from?* (If not home, workplace or shops)
- Which of the following best describes why you were there?*
- Where were you headed for?* (If not home, workplace or shops)
- Which of the following best describes why you were going there?*
- Including yourself, how many people were in your vehicle at the time of the accident/near-accident?*
- At what time of day did you start the trip?*
- Did you share the driving with another driver during the trip?*
- On that day, not counting the times when you may have swapped driving with another driver, did you have any breaks form driving after you started the trip?*
- How many of these breaks did you have?*
- On average, roughly how long were these breaks?*
- Thinking of the time just before the accident/near-accident, had you noticed that you felt tired or fatigued?*
- Had you done anything during the trip to try and stop you feeling tired or fatigued? (If yes) What did you do? (Probe)*
- Drivers have said different things as to why they do not stop when they are feeling tired. Thinking about the time just before the accident/near-accident that you had, why had you not stopped driving?
- Did you have any alcohol in the five hours before the trip, or during the trip?*
- Did you have a full night's rest the night before your trip?*
- Thinking about the day of the accident/near-accident, roughly how long had you driven before it happened?

References

Cameron, C., 1973, Fatigue and driving – a theoretical analysis, *Australian Road Research*, **5**, 23–37.

Linklater, D.R., 1978, Traffic safety and the long distance truck driver, Research Report 8/78, NSW: Traffic Accident Research Unit, Department of Motor Transport.

Road Safety Bureau, 1992, *Driver Fatigue – A Survey of Northern Region Drivers' Attitudes and Reported Accident Behaviour,* Road Safety Bureau Consultant Report CR 9/92, Roads and Traffic Authority of NSW.

12 Motor vehicle deceleration indicators

Pasha Parpia

12.1 Introduction

It has been estimated that road accidents cost £4000 million in the UK in 1987; on average, every two hours someone was killed (4380 persons), and every two minutes someone was severely injured (263 000 persons) (*Source*: BBC TV Programme, *Tomorrow's World*, 15.5.1989). It is reported that there are 6 000 000 accidents per year in the USA between two vehicles facing in the same direction (OECD, 1971).

It is widely documented that rear-end collisions between motor vehicles form a considerable proportion of all accidents. Of all accidents in the UK in 1974 involving two or more vehicles, 29 per cent occurred when they were travelling in the same direction. Of these, 23 000 colliding vehicles in accidents involving personal injury were struck from behind (HMSO, 1974). The problem is, clearly, not confined to the UK – O'Day *et al*. (1975) state that 20 per cent of all vehicles damaged in the USA are involved in rear-end collisions. The proportion is similar in other countries and independent of the year (New Zealand Motor Accident Statistics 1968–69; Andreassend, 1976; Mortimer, 1969). The fatal or serious rear-end accident rate is twice as high during hours of darkness compared to daylight hours. The most common form of rear-end collision, however, involves a moving vehicle running into the rear of a parked car (Cole *et al*., 1977).

Although it has been suggested that drivers should attempt to maintain sufficient headway between moving vehicles – a two-second temporal gap is recommended by Hills (1980) – it is not observed by many drivers. Some 11 per cent of drivers leave themselves a gap of half a second, or less. He suggests that this may be because these drivers have learned from experience that emergencies arise rarely; and, indeed, it is as problematic to gauge such a time interval while driving as it is to assess the following distances at the different driving speeds recommended by, for instance, the Highway Code. However, the incidence of accidents may be reduced by use of more efficient signalling systems.

Road vehicle density

A significant aspect of current trends in road building, coupled with increases in the number of motor vehicles registered, is that the density of vehicles (number of vehicles per kilometre of road length) has increased substantially in the past 50 years. UK road transport statistics (HMSO, 1973; 1989) show that the total number of vehicles registered has increased from 3.1 million in 1939 to 23.3 million in 1988, representing a 7.4-fold increase, while the length of roads has increased by only 22 per cent in this time (from 290 000 km to 354 000 km). This amounts to roughly a six-fold increase in vehicle density (from 10.8 vehicles/km to 65.8 vehicles/km) in this period. Current estimates suggest that there will be an approximately linear increase in vehicle density at the rate of about one additional vehicle per km per year for the next 25 years (HMSO, 1989; British Road Federation Statistics, 1985). Not taken into account, were increases in the number of lanes of the roads in working out the vehicle densities stated above.

Clearly, it is usually cheaper to widen roads to accommodate a larger number of vehicles than to build new ones; particularly in rural areas. The UK Ministry of Transport announced that road traffic (measured in vehicle-km/year) was expected to increase by 140 per cent between 1989 and 2025, and financial provisions would be made to widen, strengthen and add to the existing highway system in Britain (HMSO, 1989). However, while a widening of the roads will allow a greater flow of traffic, it can also increase the risks of multiple pile-ups that can involve several carriageways; and, indeed, accidents involving over 100 motor vehicles have been reported.

> Munich (AP) – A chain-reaction pile-up involving 115 vehicles on autobahn left 26 people injured yesterday, four of them seriously, police here said. The police blamed the collisions on drivers travelling too fast during heavy rain. The pile-up, near Allerhausen about 28 miles north of here, created a traffic jam 12 miles long on the autobahn between Nuremberg and Munich.
>
> (Source: *The Times*, 5.9.1989.)

Motor vehicle designers have largely responded to the need for greater road safety through hardware solutions directed at minimizing personal injury to the occupants of vehicles. The improving of signalling systems, however, provides an inexpensive means of lowering personal (and material) injury through the prevention of accidents through better communication. Better signalling systems have the potential to improve matters; particularly under adverse weather conditions compounded with driver fatigue, when perceptual and cognitive errors are more likely to occur.

Braking behaviour

Probst (1986) documents the times taken for the stages a driver goes through in bringing a moving vehicle successfully to a standstill in an emergency braking manoeuvre.

- A: The time required for the perception of the need to apply the brakes

    This time was determined by observing the driver's initial saccadic eye movements, culminating in the fixation of the eyes on the danger (55–300 ms).

- B: The time required for recognizing the object/danger (184 ms average)

- C: The time required for making the decision to apply the brakes (170–310 ms)

- D: The time for the foot to release the throttle pedal and move to the brake pedal (190 ms)

- E: The time for the brakes to grab (50 ms)

- F: The time to compress the brake cylinder until maximum brake response is obtained (170 ms)

- G: The time for the vehicle to come to a standstill (clearly, a variable dependent on many factors).

It must be noted in applying the times given above that during experimental investigations such as this one, the driver is already aware of what is being investigated. In actual driving situations these times, particularly the ones concerned with becoming alerted to the need for braking, are likely to be longer.

Lee (1976) notes that, in general, a driver does not initiate braking as soon as he or she realizes that the vehicle is on a collision course, for to do so would unnecessarily slow his progress. Rather, the driver adjusts braking according to his or her assessment of the overall situation; for instance, anticipating that an actual collision is unlikely to eventuate.

For the present purposes, braking behaviour may be regarded as consisting of two essential steps:

1. the recognition of the need for applying the brakes; and
2. the application of the brakes in a controlled manner (including the initiation of the many physical processes it entails).

Phase 1, then, comprise Probst's steps A–C, and is referred to, broadly, herein as 'thinking time', and step 2 the 'braking time'.

Cole *et al.* (1977) have reviewed in some detail the means previously suggested in the literature for reducing both thinking and braking times, and observe that improvements could be made to reduce both phases 1 and 2 mentioned above.

Reduction in thinking time

There is evidence that the presence of standard stop lamps on motor vehicles reduces thinking time from approximately two seconds to one second for both mild braking and medium braking (0.14 g and 0.2 g respectively) (Rockwell and Treiterer, 1966), as well as in cases where there was need for severe braking (0.3–0.8 g) (Rosemann, 1976). In the latter report, a 25 per cent reduction in reaction times was claimed when stop lamps were functioning compared to when they were not. These last results were obtained using a simulator, and it must be noted that discrepancies have been reported by Probst (1986) between results obtained for reaction times for the detection of a closing of distance between vehicles from studies employing simulators, that provide only visual information, and actual road trials. Road trials in these experiments consistently revealed reaction times that were longer by a factor of 1.5–3.0 than in simulations of the same situations. That improvements in reaction times to braking are afforded by stop-lamps is, however, not in question.

Cole *et al.* (1977) note, however, that

> the brake signal is feeble and [its] coding mode is inappropriate. [That there is a] need for higher intensities and separate function. When the brake signal is operated the signal emitted is not a compelling one ... [That] in daylight ... it cannot be expected to attract attention reliably. It is more likely that the following driver will respond to increasing angular size of the lead car rather than the onset of the brake signal.

They go on to observe that even though the onset of the illumination of the stop lamp is readily detected above that of the tail light at night, if for some reason the initial onset is not detected, the following driver cannot be certain that the stop signal is showing. They suggest that a better system would be one where there is a redundancy of function through, for instance, a spatial separation of the stop lamp from the tail lamp to reduce or remove this ambiguity. With regard to the use of a different colour to provide redundancy, they cite the OECD report (1971) which states that 'colour as a primary coding dimension is not a good choice'. Cole *et al.* (1977) also point out that having only two stop lamps on motor vehicles affords only a limited hardware redundancy in the event of lamp failure. They observe that the only mandatory coding variable (in 1977) was intensity. The others, they suggest, that may be considered are area and spatial separation.

High-level stop lamps

There was a considerable interest in rear lighting and signalling in the USA in the late 1970s and early 1980s. For a review of the literature, see Meatyard et al. (1988). One of the innovations of this period was the high-level stop lamp. In brief, some 5000 cars were fitted with high-level stop lamps and a similar number of control vehicles were monitored while travelling a distance of 50 million miles. It was found that the cars fitted with the device had approximately a 54 per cent reduction in rear-end shunts compared to the control vehicles. As of 1 September 1988, high-level stop lamps have been required by law in the USA on all new cars. In the UK, the British School of Motoring fitted these devices to all of their 2300 cars, and reported a 25 per cent reduction in rear-end accidents (Source: *The Times*, 7.7.1989).

It is difficult to assess, however, the extent to which these dramatic reductions in accident rates may be an artefact attributable to a novelty effect, which will persist while only a small proportion of road vehicles are equipped with the devices. It must be noted that many drivers complain of distraction and dazzle from high-level stop lamps, and in recognition of this, the UK Department of Transport proposed in their 1985 Draft: Road Vehicle Lighting Regulations (HMSO, 1988) that high-level stop lamps be only half as bright as the European specification for standard stop lamps. In response to complaints by drivers, it was reported in the late 1980s, that the German Transport Authority did not permit the fitting of high-level stop lamps because they considered that the eye-level light is a dazzling irritant – not a safety aid (Source: *The Times*, 7.7.1989).

It is interesting to note from the work of Sivak et al. (1984) on driver eye-fixations while following other vehicles, that drivers tend to concentrate their attention in the area of the rear screen of the preceding vehicle, and that only a little attention is paid to the region where the ordinary low-mounted brake and tail-lights are generally located. Clearly, such behaviour allows drivers to not only pay attention to the car directly ahead, but also obtain information *through* its front and rear screens about the scene ahead (and anticipate the need for braking, particularly if all preceding vehicles are fitted with high-level stop lamps, which are more likely to be visible than standard brake-lights).

Thus, in terms of alerting following vehicles to the need for a braking response, high-level stop lamps may be regarded as an improvement on standard stop lamps. They provide greater redundancy in the event of brake-lamp failure; they provide spatial redundancy as well as overall intensity redundancy. They do not, however, provide any additional information in terms of the intensity of braking of preceding vehicles, and it is not clear if they may be justified because of the nuisance and distraction they create; particularly if a more informative system might be devised. Recent articles evaluating high-level stop lamps have been presented also by Fowkes and Storey (1993) and Akerboom et al. (1993).

Reduction in braking time

Lee (1976) notes that drivers base their judgement of how to control the application of their brakes on their visual assessment of their rate of closing and the distance between their vehicle and the vehicle ahead. The application of the brakes by a leading vehicle is usually conveyed by its stop lamps, and, until sufficient time has elapsed for a driver to gauge the deceleration of the vehicle ahead, he or she must hold back from applying too great a pressure on the brake pedal if an unnecessary loss of speed, or, for that matter, a rear-end shunt to his or her own vehicle, is to be avoided. As mentioned above, both standard and high-level stop lamps can only inform following drivers that the vehicle(s) ahead has its brakes applied. There is no information conveyed through them about when to brake, how hard to brake, or for that matter whether braking is at all necessary.

A better system of signalling would be one in which the severity of deceleration is displayed in a mode that was quantitative and directly useful. A requirement of such a system would be that following drivers could make a comparison of the deceleration signals of the leading vehicles with a feedback display of their own vehicle's deceleration, without averting their gaze from the primary task domain (Wierwille, 1993), to assist in controlling their own braking.

The successful communication of such information could, in principle, substantially alter the braking behaviour of a following driver; particularly in the event of the immediate need for full-power braking.

Previous signalling systems

A number of signalling systems have been proposed in both the academic and patents literature directed at communicating the magnitude of deceleration to following drivers.

Engine-deceleration displays

Systems have been proposed for the signalling of the onset of engine deceleration through:

1. a throttle pedal actuated switch (Crosley and Allen, 1966; Rockwell and Treiterer, 1966; Hanner, 1967; Mortimer, 1971; Gorman, 1972), and
2. an inertial decelerometer (Rutley and Mace, 1969).

Such systems can be of considerable value in reducing the response times for braking by following drivers, as has been demonstrated by Rockwell and Treiterer (1966) and Rutley and Mace (1969). But, although throttle release precedes the application of the brakes – Cole *et al.* (1977) note that 350–750 ms may generally elapse in the interim – the resulting decelerations are typically small, and given that it is often *not* followed by the application of the brakes, the routine signalling of throttle release, or small decelerations – 0.05 g in the experiments of Rutley and Mace (1969) – by all road vehicles to following drivers is likely to lead to the generation of a lot of superfluous information. A distinction that, perhaps, needs to be made is the one between inadvertent or incidental decelerations and intentional decelerations. Throttle release when gear change is taking place would be an example of the former, while the use of engine braking to increase the following distance when a potentially dangerous situation appears to be developing ahead, is an example of the latter. In both cases, deceleration results and must be taken into account by a following motorist. However, in the latter example, of intentional deceleration, the activation of a signal indicating its onset could be of some value to following traffic. Gorman's (1972) system is interesting in that it attempts to filter information output through not actuating the signalling system until the foot is lifted clear off the throttle pedal.

Direct display of deceleration

Rockwell and Treiterer's (1966) AID system had a row of ten red lights which came on progressively with deceleration, so that at the start of the deceleration, the outermost lights came on, and as deceleration increased the inner lights also came on. They found that the mean response times for the initiating of braking by a following driver using this arrangement were not shorter than for conventional stop lamps. Rutley and Mace (1969) investigated a deceleration indication system which was similar to the one proposed by Rockwell and Treiterer (1966), except that a row of seven equally spaced red lights positioned in the place of the rear bumper was used. They came on in the reverse sequence to indicate deceleration severity to the one used by Rockwell and Treiterer: the innermost came on for mild decelerations while the outer lights were also switched on for higher decelerations. In this arrangement, they expected to utilize the well-known response in humans of avoidance behaviour on becoming alerted to a dilating visual field (Bower et al., 1970). It was hoped that the deceleration signal would create the illusion of an expansion of the rear of the leading vehicle. They, however, found that their system induced a small but not statistically significant reduction in the response times for braking in the driver of a following vehicle. The tests were restricted to trials on a private test track, and it would appear that all the experiments were conducted during daylight hours under conditions of good visibility. To attempt to address the problem of recreating a real-world driving situation, with its many distractions, the driver in the following car was required to perform a subsidiary task to divert his concentration from the driving.

It is worth remarking that for there to be an illusion of an expansion of the rear end of the car, the cues by which the width of the car is defined should appear to move outwards. In daylight, the lateral edges of the vehicle would have been clearly visible to the following driver. As such, it seems unlikely that the illusion of an expansion that they sought could be induced under these circumstances.

Riley (1964) proposed a system identical in its display scheme to that of Rutley and Mace (1969). However, there is no report of the efficacy of the device.

Voevodsky (1967) proposed the fitting of a centre-mounted amber light on the rear of vehicles which flashed when the brakes were operated. The rate of flashing increased exponentially with deceleration, reaching a maximum rate of 7.5 Hz at 0.5 g. The system was evaluated in road trials (Voevodsky, 1974) by having them fitted to 343 San Francisco taxi cabs. He reported that the average collision rate of 8.91 collisions per million miles over 7.2 million miles for control vehicles was reduced to 3.51 collisions per million miles for the cabs fitted with the flashing device (a 39 per cent reduction).

Mortimer (1981), however, reports that the rear-end collision rate is approximately the same for cases where the amber light is

(a) flashing,
(b) flashing at a rate that is proportional to the deceleration, and
(c) steady burning.

These results suggest that information about the magnitude of deceleration was not being used by the drivers in the study conducted by Voevodsky (1974) (if the difference between the linear dependence in flash frequency with deceleration, used by Mortimer, and the exponential change in flash frequency with deceleration used by Voevodsky may be neglected).

Mortimer and Kupek (1983) have evaluated the number of different flash-rate regimes that could be differentiated in the range 1–7 Hz by human subjects in connection with the system proposed by Voevodsky (1967). They report that only three flash rates could be discriminated (e.g. 1.0 Hz, 2.6 Hz and 6.8 Hz) using absolute judgement (rather than through comparison).

There are, however, inherent limitations associated with the use of a flashing light as a signalling system. The upper bound on the usable flash rate is, clearly, determined by the human being's capacity to perceive the flashes separately to assess their frequency. This capacity depends on number of factors (Murch, 1973), but about 15 Hz would be the maximum usable frequency. The lower bound would be determined by whether or not the time taken to decode the information from the flashes is short enough for the signal to be useful. At a flash rate of 10 Hz in the driving situation, at least two or three flashes would have to have been emitted before a following driver might be expected to apprehend and gauge the flashing frequency. As such, at best 200–300 ms would have elapsed of the driver's thinking time before he or she could know the flashing frequency realm in which to place the signal. At lower flash rates, this interval would increase. These considerations place limitations on Voevodsky's system for the display of deceleration. Clearly, a system which could convey information about the magnitude of decelerations more quickly would be preferable.

Lee (1976) proposed a two-level 'imperative' brake light signal, activated by a transducer in the brake line, and which became illuminated when the hydraulic pressure in the line exceeded a predetermined level. The imperative signal would remain on until cancelled by pressure on the throttle pedal. He suggested that the signal could take the form of an L-shaped bar of red light (and its mirror reflection) positioned at each side of the rear of the vehicle. Conventional 'mild-warning' brake lights would be placed at the corners of the L. In emergency braking, the imperative brake lights would be illuminated automatically, and in normal stopping the driver could activate them by brief hard pressure on the brake pedal. In both cases, the lights would remain on till the car stopped, and would only go off when the car moved off again. It is not known if such a system was implemented. Use of a transducer in the brake line to infer decelerations will, however, result in different onset deceleration values for the activation of the imperative brake signal, depending on the (variable) laden mass of the vehicle, and would be difficult to standardize. Use of an inertial decelerometer would, of course, obviate this problem.

Schimmel (1978) suggested an arrangement similar to the one described by Rockwell and Treiterer (1966), where a row of lights on the rear of the vehicle is turned on at its widest for mild braking and fill in a band of illumination with increasing deceleration. The system was activated by brake pedal travel. There is no report of an evaluation of the system through experimental trials. Once again, this means of activating the signalling system is inappropriate if the onsets of indications are to occur at predefined deceleration values because brake pedal travel is unlikely to remain constant with wear and so on.

Athalye (1985) suggested a deceleration indication system comprising a number of different coloured lights to indicate severity, activated by an inertial decelerometer, which was compensated for deviations of the inclination of the vehicle from the horizontal. He provides no report of an evaluation of the system.

12.2 The proposed system

The system presented below attempts to address the need for reducing both the thinking times and the braking times. It has been described previously in Parpia (1993), and has been implemented on a vehicle in the UK. A number of details of

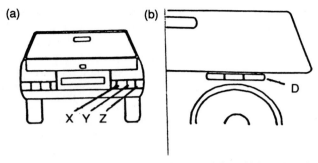

Figure 12.1 (a) The position of lamps X, Y and Z at the rear of the vehicle and (b) the lamps D on the dashboard.

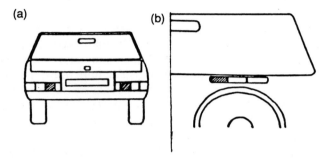

Figure 12.2 Mild deceleration: (a) the tail lights; and (b) the dashboard.

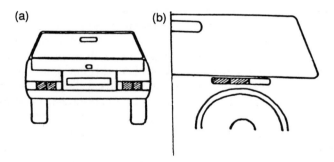

Figure 12.3 Increasing deceleration: (a) the Y light is also on; and (b) shown on the dashboard.

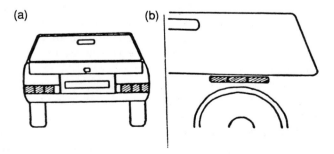

Figure 12.4 Severe deceleration: (a) all lamps are on at the rear; and (b) on the dashboard.

the arrangement still remain to be determined. However, its basic principles of function are outlined herein.

Figures 12.1a and 12.1b illustrate the system in its un-activated state. The rear of the vehicle has an expanding array of stop lamps (marked X, Y, and Z), which are activated in pairs by an inertial decelerometer. In addition, a 'self-display' system (D) is made available for the driver of the vehicle; this informs the driver about the status of the signals at the rear of the vehicle, and the driver thereby, has access to information about his or her own decelerations. The innermost lamps, X, act as both presence lights as well as a part of the deceleration indication system. For mild decelerations, only the innermost pair of lights (X) are illuminated with enhanced intensity (Figure 12.2a). As deceleration is increased, the next pair (Y) are also similarly illuminated (Figure 12.3a). For severe decelerations, all the lamps are illuminated (with equal intensity) (Figure 12.4a). The corresponding states of activation of the 'self-display' system, D, are presented in Figures 12.2b, 12.3b and 12.4b, respectively.

The signalling system proposed addresses a number of the requirements for a brake-light display system. It is perhaps best to illustrate these while assuming that they are being observed at night-time in moderately foggy conditions in a rural setting, by a following driver, whose vehicle is also equipped with the same arrangement for signalling decelerations.

Under these conditions the apparent width of the car ahead is determined principally by its presence lights. When the car ahead is put through an emergency braking manoeuvre, the presence lights initially become brighter (as in the case of a standard combined presence/brake-light indication of the same situation). As the deceleration of the car ahead increases rather rapidly in this manoeuvre, so also does the amount of light emitted by its deceleration indicators as more lamps are turned on by the inertial decelerometer. There is simultaneously an expansion of the area of illumination, which is arranged to provoke the innate human propensity to apprehend an approaching object, conveyed by a temporal dilation of the visual field (Gibson, 1958; Carel, 1961). This arrangement has a strong attention-gaining quality, if the illusion is not broken by strong contradictory cues, which are more likely to exist when visibility is better. There are, thus, two separate and powerful cues to alert the following driver's attention to the need for braking: enhanced brightness of illumination, and looming. If the illusion of an expansion is perceived powerfully, there is a risk that it may cause alarm in some drivers, which might require a period for recovery, and reaction times for braking might be impaired. Such an effect might be reduced with a greater familiarity with the display system, when times for the recognition of the danger (Probst's stage B) might be expected to be shorter than for vehicles not fitted with the proposed system. The following driver is at once informed that the vehicle ahead is decelerating, and that this deceleration is severe. The following driver applies the brakes as soon as possible, and perceives in his or her near peripheral vision the 'self-display' being activated. While keeping foveal vision fixed on the vehicle ahead, the following driver can rapidly increase braking pressure to a suitable level, without needing to wait for such cues as would develop due to a closing of the distance between the two vehicles.

In this system, quantitative information about the severity of deceleration of both vehicles are encoded simply in the number of lamp-pairs illuminated. The driver is simultaneously alerted to the need for braking and the magnitude of deceleration of the lead vehicle. Thereafter, with the guidance of the 'self-display', along with the usual, now impaired, cues, because of poor visibility, the following driver is informed of his or her braking performance relative to that of the vehicle ahead.

Location of the 'self-display'

It is, clearly, of considerable importance that information from the driver's 'self-display' system be accessible to the driver during the controlling of braking without the need to avert his or her gaze from the rapidly decelerating vehicle ahead. The task of reading information from the 'self-display' is not an overly demanding one, as it requires only the detection of the presence or absence of illuminated bars, whose number indicate the level of deceleration. Nevertheless, some consideration needs to be given to the size and location of this display.

Cohen and Hirsig (1991) note that the time required to pick up the visual information intended is shortest when fixating an object, and increases monotonically with increasing eccentricity of the retinal image. In addition, near-to-far and far-to-near accommodations of the eye in the car driving situation requires 100–400 ms (Hartmann, 1970). It is noted, however, that in normal driving the visual demands are not overly severe: for purposes of steering along a straight road, foveal fixations deviating by 4–5° from the path of travelling is adequate for successful performance (Bhise and Rockwell, 1971).

But, it is observed by Miura (1986), through an eye-movement study in actual driving, that there is both a decrease (narrowing) of peripheral vision performance and an increase in reaction times as the driving situation becomes more demanding. This finding is reinforced by those of Labiale (1993), who observed that the preferred location of a motor vehicle warning light to produce the shortest response time for its detection was when it was located closest to the domain of the functional attention field for the primary task (the road ahead).

It might be worth considering the use of a head-up presentation of the 'self-display' focused at an appropriate depth and location on the front windscreen. Clearly, attention would have to be paid to any interference caused by this to the driver's overall performance in evaluating this means of presentation.

In conclusion, it seems that the individual luminous elements of the 'self-display' should be large and located as close as possible to the domain of the primary task (Wierwille, 1993) in such a way that it produces the minimum of distraction during normal driving, and is readily accessible to the driver's near-peripheral vision. This could be achieved if, for instance, the 'self-display' was located on the dashboard, near the lower edge of the front windshield, slightly to the passenger's side of the steering wheel, so that the the wheel did not obscure it from the driver. Its precise dimensions and location have yet to be determined, but may not be too critical within the suggested bounds.

Location of the rear-end deceleration display

There are two locations for the rear-end display that merit consideration:

1. incorporated among the existing standard tail-lights and indicators at a low level; and

2. fixed at eye-level.

The virtue of the first is that the illusion of a dilating visual array is likely to be better served if the presence lights can appear to move outwards. Also, being closer to the previously established 'arena' for the display of signals, it is likely to be more useful. Also, in this location, it will create less distraction. A further advantage of such a location for the display is that the angular distance between it and the suggested location of driver's 'self-display' is smaller, and would facilitate comparisons.

The advantages of the second choice are essentially the ones attributed to high-level stop lamps: signals are provided in the domain where the driver's eyes generally dwell to gain information both about the movements of the car ahead as well as of the general flow of traffic and road conditions through the rear and front windshields of preceding vehicles. In this location, although there may be a loss of the attention-gaining quality of the signal system because of the loss of apparent outward movement from the standard presence lights, they would retain all the advantages of gaining attention that high-level stop lamps have (and more). But, in this position, there are also the risks of masking the view of the road ahead in conditions where the brightness of the deceleration indicators is greater than the image of the roadway beyond, discernible through the screens of the vehicle ahead. Also, in this location, they are more likely to cause annoyance through distraction and dazzle in routine usage.

Another possible location, that is, perhaps, less likely to find favour because of its novelty, is the top edge of the vehicle: between the top of the rear screen and the roof of the car, where presence lights are sometimes located. In this position, high-level presence lamps might once again represent the width of the vehicle at night-time, and be seen to 'expand' with the onset of rapid deceleration.

However, this location appears to be a poor choice because, unlike the site of the standard signals at the rear-end, which lies between the primary functional attention field and the secondary attention field (of the car's instrument panel), the top edge would require a diversion of foveal vision in the opposite direction for the indicators to come under scrutiny.

12.3 Discussion

A system has been proposed that addresses the need for reducing both thinking time and braking time through the implementation of an expanding array of stop lamps. These are designed to alert following drivers to the need for braking by a number of redundant signals known to elicit attention. Furthermore, through the calibration of decelerations into predetermined intervals, the actual deceleration is indicated unambiguously through the number of lamp-pairs illuminated. The provision of a 'self-display', that informs the driver of his or her own deceleration, allows a comparison to be made of the driver's own deceleration with that of leading vehicle(s) to provide a guide for the controlling of braking. The expanding array of indicators is similar in its function to those proposed by Riley (1964) and Rutley and Mace (1969). The provision of the 'self-display' is an innovation.

However, there is no substantial evidence from driving trials thus far reported that the provision of deceleration indicators can reduce the incidence of rear-end shunts. Equally, there is no convincing argument on theoretical grounds that they cannot be of use. The intuition is that there are many situations in which a motorist would feel more secure to have signalled the magnitude of the deceleration of a vehicle ahead. It is conjectured that such signals, as have been proposed herein, would take motorists a little while to become used to; to come to have confirmed the relationship between the signals and the resulting decelerations; also to come to rely on the worth of the 'self-display' for establishing a match in decelerations. With the development of an ease in using the system, the response to signals of rapid deceleration would operate with a reduced cognitive load. In braking behaviour, the steps pertaining to eye fixation on a high deceleration signal; to recognizing it, and the step of making the decision to apply the brakes are all likely to be shorter, particularly in adverse conditions. Furthermore, the subsequent control of braking is likely to be better handled by the driver with the aid of the 'self-display'. These intuitions have yet to be evaluated.

Rutley and Mace (1969) used a similar system to the one proposed herein to give warning of engine deceleration. Given that the proposed system is activated also by an inertial decelerometer, its threshold for initial onset may be adjusted to respond also to appreciable engine decelerations. These indications can be inadvertent in the course of, say, poor clutch–throttle synchrony on changing gears, or they may be due to the onset of intentional engine deceleration preceding the application of brakes.

This raises the interesting question concerning the desirability of signalling inadvertent decelerations. One might insist that signals be restricted to communicate only the driver's intentions. For instance, a driver anticipating a hazard might remove his foot from the throttle pedal and place it on the brake pedal, ready to press it should the need to brake suddenly arise. If this resulted in the activation of the stop lamps, such information might be more pertinent to a following motorist (if he or she knew that only intentional signals were allowed to be communicated) than a 'passive' signal of engine deceleration which could result from allowing the inertial decelerometer to signal the onset of engine deceleration. Schedule 12 of UK Draft Road Vehicle Lighting Regulations (HMSO, 1988) requires that the stop lamps 'shall be operated by the application of the service braking system'. Presumably, this recognizes the importance of the communication of intention.

As remarked earlier, a driver can usually obtain information about the roadway ahead by looking through the windscreens of preceding vehicles, thanks to the development of larger rear screens. This facility is not afforded to someone who has a vehicle like a closed lorry or a large omnibus either directly ahead or even remotely ahead, since all sight of what lies beyond is obscured by such vehicles. The fitting of deceleration indication system to them would be of particular use to following motorists as a source of early warning of rapid deceleration.

The array of red lamps of the deceleration indication system provide a signalling facility that might be used also for indicating vehicle presence while parked on the highway. For instance, the lamps could be put into a state where they display in a random sequence short pulses of illumination from the lower intensity filaments of the stop lamps to indicate parked presence. Having a plurality of lamps flashing in a random sequence would avoid their flashes being confused with turn indication (turn indicators in the USA are fitted with red filters). However, any implementation of this proposal would first require a careful evaluation of its benefits in relation to costs through the distraction it might create for other road users.

The suggested use of red lamps for the deceleration indicators are in sympathy with the need for conformity of any new system with the one already established for signalling braking.

There have been for some time calls for the standardizing of rear-end signal systems to enhance reaction times through a reduction of ambiguity (OECD, 1971; Cole *et al.*, 1977). If the system proposed merits further consideration, it might be incorporated within a standard design of a rear-end signalling array. It may be appropriate then to reconsider the debate of whether an eye-level display for deceleration or its incorporation into the rear-end display is preferable.

For the proposed system to be widely applicable, there would be need for the establishment of suitable deceleration values for the onset of the different lamps that comprise the system, so that meaningful matches may be made across vehicles.

Acknowledgements

The author is grateful to his many colleagues and friends who have given their time and generously made accessible their insights into the problems of signalling of motor vehicle deceleration. In this respect, he would like to particularly mention Charles North, Alistair Bray, Jasper Taylor, Jim Stone, David Young and Rod Lauder, all, at one time or another at Sussex University.

References

Akerboom, S.P., Kruysse, H.W. and Heij, W.L., 1993, Rear light configurations: the removal of ambiguity by a third brake light, in Gales, A.G., Brown, I.D., Haselgrave, C.M., Kruysse, H.W. and Taylor, S.P. (Eds) *Vision in Vehicles-IV*, Amsterdam: Elsevier Science Publishers, pp. 129–38.

Andreassend, D.C., 1976, Vehicle conspicuity at night, *Proc. 8th Australian Road Res. Conf.*, **8**(5), 26.

Athalye, R.G., 1985, Brake light signal system for a motor vehicle, Patent No. US 4667177, US Patent Office.

Bhise, V.D. and Rockwell, T.H., 1971, Role of peripheral vision and time sharing in driving, *Proc. Am. Ass. of Automotive Medicine*, Colorado, Colorado Springs.

Bower, T.G.R., Broughton, J.M. and Moore, M.K., 1970, Infant response to approaching objects: an indicator of response to distal variables, *Perception and Psychophysics*, **9**, 193–6.

British Road Federation, 1985, *Basic Road Statistics*, London.

Carel, W.L., 1961, Visual factors in the contact analog, General Electric Electronics Centre Publication, Cornell University Press, No. R6ELC60.

Cohen, A.S. and Hirsig, R., 1991, The role of foveal vision in the process of information input, *Proc. Third Int. Conf. on Vision in Vehicles-III, Aachen*, Amsterdam: Elsevier Science Publishers, pp. 153–60.

Cole, B.L., Dain, S.J. and Fisher, A.J., 1977, Study of motor vehicle signalling systems, Commonwealth of Australia Department of Transport Report on Project 73/844.

Crosley, J. and Allen, M.J., 1966, Automobile brake light effectiveness: an evaluation of high placement and accelerator switching, *Amer. J. Optom. Arch. Amer. Acad Optom.*, **43**, 299–304.

Fowkes, M. and Storey, H.S., 1993, Evaluation of rear lighting equipment, in Gales, A.G., Brown, I.D., Haselgave, C.M., Kruysse, H.W. and Taylor, S.P. (Eds) *Vision in Vehicles-IV*, Amsterdam: Elsevier Science Publishers, pp. 119–28.

Gibson, J.J., 1958, Visually controlled locomotion and visual orientation in animals, *British Journal of Psychology*, **49**, 182–94.

Gorman, J.M., 1972, Visual warning apparatus for vehicles, Specifications of Australia, Phillips, Ormonde and Fitzpatrick, Patent Attorneys.

Hanner, M.T., 1967, Vehicle control light system, Patent No. GB 1177969, The Patent Office, London.

Hartmann, E., 1970, Sehn und Beleuchtung am Arbeitsplatz, Munchen: Goldmann.

Hills, B.L., 1980, Vision, visibility and perception in driving, *Perception*, **9**, 183–216.

HMSO, 1973, *UK Department of the Environment Statistics*.

HMSO, 1974, *Road Accidents in Great Britain*, UK Department of Transport Road Accident Statistics.

HMSO, 1988, *Draft: Road Vehicle Lighting Regulations*, UK Department of Transport.

HMSO, 1989, *Roads to Prosperity*, Department of Transport Statistics.

Labiale, G., 1993, Visual detection of in-car warning lamps as a function of their position, *Proc. 4th Int. Conf. on Vision in Vehicles, Liden*, Amsterdam: Elsevier Science Publishers, pp. 291–9.

Lee, D.N., 1976, A theory of visual control of braking based on information about time-to-collision, *Perception*, **5**, 437–59.

Meatyard, A.G., Fowkes, M. and Wall, J.G., 1988, An investigation into rear lighting arrangements for cars, *Int. J. Vehicle Design*, **9**, 471–80.

Miura, T., 1986, Coping with situational demands: a study of eye movements and peripheral vision performance, *Proc. Conf. on Vision in Vehicles, Nottingham, UK, 1985*, Amsterdam: Elsevier Science Publishers, pp. 205–16.

Morrison, W.J. and Morrison, G.W., 1938, Improvements in or relating to speed indicators for vehicles, Patent No. GB 492328, The Patent Office, London.

Mortimer, R.G., 1969, Research in automotive rear lighting and signalling systems, G.M. Corp, Engineering Publication 3303.

Mortimer, R.G., 1971, The value of an accelerator release signal, *Human Factors*, **13**, 481–6.

Mortimer, R.G., 1981, Effects of braking deceleration signals on rear-end collisions, *Proc. Ann. Conf. Amer. Assoc. for Automobile Medicine*, pp. 369–80.

Mortimer, R.G. and Kupec, J.D., 1983, Scaling of flash rate for deceleration signal, *Human Factors*, **25**, 313–18.

Murch, G.M., 1973, *Visual and auditory perception*, New York: Bobbs-Merrill Co. Inc, pp. 63–113.

New Zealand Ministry of Transport, Motor Accident Statistics 1968–69.

O'Day, J., Filkins, L.D., Compton, C.P. and Lawson, T.E., 1975, *Rear impacted vehicle collision: frequency and casualty patterns*, Motor Vehicle Manufacturers Association, Highway Safety Research Institute, University of Michigan.

OECD, 1971, *Lighting, visibility and accidents*, Report prepared by an OECD Research Group Organisation for Economic Co-operation and Development, 4.

Parpia, D.Y., 1993, Vehicle deceleration indicator system, Patent No. GB 22295323, The Patent Office, London.

Probst, T., 1986, Detection of changes in headway, *Proc. Conf. on Vision in Vehicles, Nottingham, UK, 1985*, Amsterdam, Elsevier Science Publishers, pp. 157–66.

Riley, A.B., 1964, Vehicle deceleration indicator, Patent No. US 3157854, US Patent Office.

Rockwell, T.H. and Treiterer, J., 1966, Sensing and communication between vehicles, Ohio State Univ., Systems Research Group, Final Report No EES 272-1 Columbus, Ohio.

Rosemann, R., 1976, Einfluss von Ruckleuchtensystem auf die Bremsreaktionszeit von Kraftfahrzengfahrern, *Proc. 18th Session CIE, London*, CIE Publication no. 36, Paris.

Rutley, K.S. and Mace, D.G.W., 1969, An evaluation of a brakelight display which indicates the severity of braking, UK Road Research Laboratory, RRL Report No. LR 287.

Schimmel, G., 1978, Bremsgradanzeigevorrichtung für ein Kraftfahrzeug, Patent No. DE 2637155, W. German Patent Office.

Sivak, M., Conn, L.S. and Olson, P.L., 1984, *Driver Eye Fixations and the Optimal Locations for Automobile Brake Light*, University of Michigan Transport Research Institute Report No UNTRI-84-29.

Voevodsky, J., 1967, Inferences from visual perception and reaction time to requisites for a collision-preventing cyberlite stop lamp, *Proc. Nat. Acad. Sci. USA*, **57**, 688–95

Voevodsky, J., 1974, Evaluation of a deceleration warning light for reducing rear-end automobile collisions, *J. Appl. Psychology*, **59**, 270–3.

Wierwille, W.W., 1993, An initial model of visual sampling of in-car displays and controls, in Gales, A.G., Brown, I.D., Haselgrave, C.M., Kruysse, H.W. and Taylor, S.P. (Eds) *Vision in Vehicles-IV*, Amsterdam: Elsevier Science Publishers, pp. 271–80.

13 An alcohol and other drug education programme for bus operators

Tanya Barrett

13.1 Introduction

Alcohol and other drugs (AOD) are an increasing problem for modern society (James, Harrison and Laughlin, 1993). With Australia having a higher level of alcohol consumption per person than other English speaking nations, this is of concern to both the general community and employers. The Health Department of Western Australia reports that, in fatal road crashes, half can be attributed to the use of alcohol. For industries who employ professional drivers, who are regularly required to work long hours and weeks, alcohol adds another concern to the other workplace risks, of driver fatigue, driver stress – due to workload and critical incident stress – as well as reduced general fitness due to the effects of shift work. With fatigue being a major concern with drivers, alcohol can lead to similar problems, tiredness, reduced alertness and attention, and compound the effects of fatigue in drivers.

As a responsible employer, industries who employ 'professional' drivers – such as bus drivers, taxi drivers and truck drivers – are facing another risk in AOD use. This risk, which affects work performance and safety, is not directly under the control of the employer. This chapter does not suggest that any of these driver groups do, or do not, use AOD in a manner which places the driver, co-workers or the public at risk. What is suggested is that as members of a community with a high level of use of at least one drug – alcohol – that there may be a need for concern and intervention. Certainly, local studies of sufficient quantity and quality are not available to support or reject either view, although for Australian industry in general, evidence is mounting that the general community alcohol consumption levels should be of concern to industry, due to hangover effects, absenteeism and possible intoxication on the job for binge drinking or workplace drinking (Blaze-Temple *et al.*, 1993).

With the growing use of drugs in Australian and other societies and the associated health costs, safety risks and other issues related to some drug usage, it is not unreasonable that there is growing interest in employer and employee groups in dealing with the issue. This is an issue of increasing concern for health and safety professionals, as it is an issue which is essentially related to an individual's non-company time, but which may 'hangover' into work time. It is a complex issue related to legislative Duty of Care, privacy, safe work practices, industrial relations, morale and rehabilitation. With AOD there are many well-documented studies on their effects on an individual's performance (Laurell, and Tornros, 1991; Blaze-Temple, 1990). It is these effects which have the potential for major impacts on transport industries and their employees.

In 1990 Transperth, the organisation providing Perth's bus services, commenced a comprehensive training programme referred to as 'Transit Ambassador' which was purchased from Canada. The programme was purchased with the aim of providing ongoing training and education for bus operators. A number of 'in-house' modules were developed to suit the package, one of which is discussed in this chapter relating to alcohol and other drug training for bus operators. The purpose of the training was to prepare Transperth employees, in particular bus operators, for the development of an AOD policy in the workplace. The aim of the educational programme was to ensure that employees had adequate information about the effects of drugs on work performance and health, so that when asked by Transperth on how the policy should be developed, informed decisions were made for themselves and the workplace. The programme also aimed to demonstrate that the issues of AOD were not simple and there was no single solution. The programme has since educated 1000 bus operators and the formal and informal feedback from bus operators is excellent, and resulted in other non-bus operator union groups requesting similar training. For this and other alcohol and drug services provided, Transperth were awarded the 1990 Alcohol Awareness Award.

The author acknowledges the fact that an educational programme is an integral rather than 'stand alone' factor in prevention and management of AOD in the workplace.

13.2 Method

Objectives

1. To provide accurate information on Transperth's Health and Welfare services which were available to employees
2. To provide accurate information regarding the effects of AOD use
3. To provide the necessary educational component to employees prior to the introduction of an AOD policy
4. To reinforce the need for healthy and safe AOD use behaviours among bus operators.

Programme design

The programme was designed by the Occupational Health Co-ordinator and the Welfare Co-ordinator at Transperth, with the assistance of Health Promotion Services of the Health Department of Western Australia, Alcohol and Drug Authority and Indrad Services, whose contribution is gratefully acknowledged.

The programme was of three hours duration, and was conducted between bus operator shifts. This in effect meant that participants drove a shift before and after the presentation. There was a one-hour lunch break in between. The programmes were presented to groups ranging in size from 15 to 20 usually, with the occasional group of 30 participants.

Participants

Participants were bus operators drawn from several depots. They took part in the programme in groups which were composed of operators from several depots. There were two criteria for eligibility. First, they had to have five years service and second, be interested in promotion to the next level of employment. A non-attendance at any of the Transit Ambassador modules meant the next pay increment would not be awarded. However, in practice, bus operators were advised that those who had major objections to attending the AOD session were able to leave without repercussions. That is, they were involved in the introduction to the programme and if following the introduction they had serious concerns or conflicts related to culture or religion, or major objections in attending for other personal reasons, there were no repercussions if they left. This was considered important, as the messages on alcohol use may have not been acceptable to some individual's religious beliefs.

Teaching and group methods
Wherever possible interactive teaching methods such as debate, chalkboard activities, brainstorming and small group discussions were used. Content included:

- interaction of alcohol and other drugs
- blood–alcohol content (BAC) and how to work it out
- safe vs healthy alcohol use – What is the difference?
- cost-benefit analysis of alcohol use
- safe prescription medication use, and driver responsibilities
- where to obtain information about illicit drugs
- Thorley's model of drug use – intoxication, heavy regular use and dependency
- model on how the individual relates to the environment and drugs
- support services and where to obtain help
- stress management without drugs
- legislative responsibilities – what do you do if you know someone is AOD affected?

The sessions were conducted by the Occupational Health Co-ordinator, with part-time assistance from the Welfare Co-ordinator. The first session was conducted for the full Executive of the bus operators' union, who provided very useful feedback and endorsed the programme. The programme was also accredited with the Transit Ambassador programme designers, as were the presenters. It meets the requirements of the Training Guarantee Scheme. The Training Guarantee Scheme was established by Government and requires industry to offer training and development to all employees, and to meet established guidelines.

Evaluation
Evaluation was mainly through confidential formal feedback from bus operators. Questions were asked on what they learnt, whether they considered the information offered was useful and appropriate, whether they were encouraged to practise or continue to practise healthy drug use and whether they would attend again. This questionnaire had previously been used on several occasions as part of larger general health promotion at Transperth, and it is considered that participants would be familiar with it. On every question the participants rated the programme favourably.

At this stage, long-term evaluation has not been conducted as the programme has not been completed. In the future, evaluation can occur through the existing Employee Attitude Survey, and any questionnaire which may be circulated prior to an AOD policy being developed.

13.3 Results and discussion

The response of bus operators attending the groups has been very encouraging, if judged by the formal confidential feedback sheets. These indicated that in all areas the majority of participants (80 per cent plus) rated the information to be of high value, that they learned a 'great deal' to a 'significant amount', were motivated to change their behaviour to a healthier style and would attend again. The programme has had excellent management and union support, with other union groups requesting this style of programme for their members. Subjectively, some of the barriers appear to have been removed for requesting assistance from in-house services, such as the health and welfare service, and attendances have slowly increased with those using the services often reporting they heard about them through the AOD Transit Ambassador programme.

Initially participants were very anxious about the groups. It appears that many were concerned about whether they would be lectured to and questioned on their consumption. The emphasis of self-evaluation of the risks, and the frank and open discussion by many, removed the myths about what would happen in the session. The attitude of 'you are responsible' and having a choice of behaviour also seemed to lessen these anxieties.

While the presenters expected many of the myths about alcohol use and 'sobering up' to become apparent, it was obvious that bus operators, in the majority, knew that these techniques did not work. That is, they understood how to drink safely and protect their lives and others (appoint a skipper, drink at home). A knowledge gap was evident in knowing about the hangover effects associated with drinking beyond the healthy standard drink limit. Few bus operators could define a standard drink. In particular, bus operators had a lack of knowledge of how much alcohol and other drugs caused co-ordination impairment, for example, and how long it took for these effects to disappear. There was considerable anxiety over new legislation, which reduced the legal level of alcohol when driving to 0.05 BAC, until an exercise on working out the time span and factors complicating the removal of alcohol from the body was demonstrated in the session.

Participation in large groups of 30, as was to be expected, resulted in less involvement in the programme, and ratings by participants were poorer than those in smaller groups. In these larger groups, it was more difficult to encourage active participation, and ensure maximum participation.

Many participants had strong views on how AOD should be handled in the workplace. While this was not completely unexpected, there was a large group in each session who were keen to see some form of drug testing introduced. Feeling was strong against those who were perceived to be placing bus operators' good names as professional drivers at risk, through AOD use. Bus operators were critical of alcohol rehabilitation schemes which were put in place through industrial relations channels, especially as they considered that rehabilitation programmes for those who had difficulty returning to work due to ill health did not appear as equitable as for those returning to work after loss of licence related to alcohol offences. That is, if drivers lost their licence due to driving under the influence of alcohol, a job was made available to them until the licence was regained. Those with ill health were not routinely offered alternative duties.

13.4 Summary and recommendations

The major recommendation is that these AOD programmes continue to be conducted in the workplace. It should be noted that AOD use is not the only factor which contributes to driver fatigue. In order to maximize any health intervention, all issues should be dealt with, rather than dealing with a major risk factor in isolation. For professional drivers, this means a review of shift structures, improved nutrition and support for other healthy lifestyle behaviours such as exercise programmes.

For Transperth this means extending the AOD programmes to all employees, with adjustments to suit age and so on. It is important that they are designed so that participants can be actively involved, and encouraged to examine their health beliefs in a safe, closed group forum. AOD use in industry manifests in driver and other employee fatigue, increased accident risk, and increased sick and Workers' Compensation leave. While industry needs to consider comprehensive management of AOD use through policy and vigilance, employee education as outlined above is essential to introduce such policies in a supportive manner, and without employee resistance. Education and consultation are two of the essential ingredients to reduce these AOD risks to industry.

Upon completion of the educational programmes, it is important that Transperth complete an AOD Policy in the near future, as employees' expectations are raised and with increased knowledge there is likely to be support rather than opposition to such a policy. Such a policy must encompass all areas of AOD use from social clubs to the board room.

The effects of these programmes must be measured. Employee attitude surveys and questionnaires which are circulated to obtain employee comment on proposed AOD policies may be a means of doing so.

Finally, these programmes offer a means to remove the barriers often seen in industry and to obtain valuable information related to a variety of topics. Such programmes can be used to fulfil the requirements of the Training Guarantee Scheme, Duty of Care, as well as be an integral part of health promotion and safety training programmes. To employees, they demonstrate an employer's concern and caring in real and tangible terms.

References

Blaze-Temple, D., 1990, Drug testing in the workplace: Overview of the issues, Symposium on *Contemporary alcohol and drug issues: implications for the workplace*, Perth.
Blaze-Temple, D., Jones, S., Keenan, S. and Yates, D., 1993, Workwell, alcohol and other drugs workplace prevention project, Curtin University, Perth.
James, R., Harrison, D. and Laughlin, D., 1993, Raising the priority of alcohol over other drug problems in a community – behavioural epidemiology versus community opinion, *Health Promotion Journal of Australia*, **1**, 49–54.
Laurell, H. and Tornros, J., 1991, Interaction effects of hypnotics and alcohol on driving performance, *J. Traffic Med.*, **1**, 9–13.

14 The effect of profile line-marking on in-vehicle noise levels

Peter Cairney

14.1 Introduction

Profile edge-lines have now been in widespread use for a number of years, particularly in Europe, and are now beginning to find applications in Australia.

There are basically three advantages associated with profile edgelines.

1. They are cheaper than conventional painted lines.
   Although more expensive to put down than paint, they have an effective life of 4–5 years compared to the six months to one year typical of painted lines. When the annual costs of providing edgelines are considered, then the advantages of the profiled lines become apparent. Because of this longer life, they are also likely to be less affected by deferred maintenance.

2. They provide improved optical guidance in wet conditions.

3. When a vehicle runs over them, audible noise is generated, warning the driver that the vehicle is running over the edgeline.

The present investigation was concerned with this last factor. There are a number of options with regard to height and spacing for the profiled edgeline markings. For example, one of the commercially available products consists of bars, laid transversely across the pavement, typically 200 mm wide and 50 mm long. For the experiment, it was decided to examine the noise generated by a range of combinations of heights and spacings, ranging from an 8-mm height with a 50-mm spacing to 4-mm height with a 450-mm spacing. As the former requires ten times as much thermoplastic material as the latter, it would appear that there may be considerable savings to be made by selecting the combination which uses the least amount of material while still delivering an adequate signal to the driver. The range selected for investigation was regarded by suppliers and practitioners as typical of the range of options that would be considered.

The only published work on the auditory effects of profile edgelines seems to be that of Soulage (1993). He reports a testing programme carried out in France which was very similar to that carried out in the present investigation, but which also incorporated user surveys. However, the types of marking tested were rather different from the ones available in Australia, so that direct comparison of the results is not possible. Being the only other work dealing with the auditory effects of profile edgeline available, it is worth describing in some detail, especially as the original article is in French.

The markings tested were either bars, with heights between 6 mm and 16 mm, with spacings between 600 mm and 1.5 m, or small buttons laid in rows, with a height of 6–10 mm and spacings ranging from 40 mm to 150 mm. Measurements were taken of the internal cabin noise, the external noise, and the vibration measured at the steering wheel, although only the first is discussed in any detail. Four vehicles were used in the tests: a motor-cycle and a truck (type unspecified) and two cars – one small and one medium. The vehicles were driven over the profile markings at 90 kph and 110 kph for the cars and motor cycle and 60 kph and 80 kph for the truck.

The data are presented in terms of the difference between each type of profile edgeline and a control section at the two test speeds. Noise levels in the cars were generally between 5 dB and 10 dB higher when driven on the edgeline; 3 dB was regarded as sufficiently loud for satisfactory performance, although no justification was put forward for this. For the cars, all markings were above this threshold, but the buttons tended to give rise to a smaller increase than did the bars.

The difference for the truck was much lower at around 1 dB, with the exception of the highest bars which were 13–16 mm high. However, the truck drivers were able to detect the change in pitch caused by the generation of sound peaking at a frequency corresponding to the rate at which the tyre contacted the marking. Although no data are presented, Soulage comments when the sound spectrum is analysed, the regular spacing of the markings results in frequencies being generated which are proportional to the travel speed.

A survey of road user opinion was carried out by having car and truck drivers follow a set route on which a variety of markings similar to those tested previously had been installed. Drivers were to run their vehicles over them and assess their auditory, vibratory and visual effects. The highest markings and the widest spacings were regarded as annoying and unsafe, and the markings regarded as most effective had heights about 8 mm and had spacings of less than 1 m. Thus the markings which corresponded best with user needs were not the ones generating most noise.

14.2 Method

Test site and markings

The test site was chosen to have the following characteristics:

1. wide sealed shoulder to allow continuous driving on edgeline at high speed;

2. level sites or slight down grade to allow easy maintenance of required speeds;

3. four-lane road to minimize disruption to other traffic;

4. adequate sight distances; and

5. close to Australian Road Research Board (ARRB).

A stretch of the South Gippsland Highway between Five Ways and Tooradin was selected. Three different heights of markings – 4 mm, 6 mm and 8 mm – were tested in combination with three different spacings – 50 mm, 250 mm, and 450 mm – making a total of nine combinations in all. Each marking was applied to the east-bound carriageway over a 300 m section. The entire length of the site was 2.7 km.

Vehicles tested and experimental procedure

Tests were carried out using three vehicles: a late-model Toyota Camry sedan, a late model Nissan Patrol long wheelbase four-wheel drive, and a six-axle articulated vehicle, the prime mover being a Kenworth Model T400. After a familiarization run with each vehicle through the site, the driver drove the vehicle along the entire length of the site with the left-hand wheels of the vehicle on the edgeline at a steady speed, while the experimenter operated the noise recording equipment. Runs were carried out at 60 kph, 80 kph and 100 kph, providing between 10 and 18 samples of the noise generated, depending upon speed. At least three runs being carried out at each speed. If in the team's judgement traffic or other factors had prevented a clear run directly on the line, the run was abandoned, and a substitute run made. Even so, a few runs were shown to be unacceptably inconsistent when analysed in the laboratory, and were discarded.

Runs were carried out with the articulated vehicle in both the laden and unladen conditions. In the laden condition, the tanker carried approximately 21 000 litres of water. In the laden condition, gross vehicle weight was approximately 38 tonnes.

Recording and analysis procedure

Noise levels inside the vehicle were recorded through a Bruel and Kjaer Model 2231 Sound Level Meter and Model 4155 microphone onto a Nagra Model IV SJ tape recorder, following calibration of the sound level meter using a Bruel and Kjaer Model 4230 Calibrator. The sound was later analysed in the laboratory using a Bruel and Kjaer Model 1625 Octave Filter in conjunction with the sound level meter and Bruel and Kjaer BZ 7103 frequency analysis software.

14.3   Results and discussion

Overall noise levels

The overall level of in-vehicle noise is presented in terms of dB(A), a weighted average of the sound pressure level in each third-octave band which is biased towards those regions of the audible range which coincide with the range of the human voice. It is widely accepted as an appropriate measure of noise levels (Kryter, 1970).

It is evident that noise levels for the Camry sedan running with one wheel on the edgeline are higher than for the control runs, generally by about 5 dB(A). It is also evident that noise levels are greater at higher speeds, with a difference of about 2–3 dB(A) between runs carried out at 60 kph and runs carried out at 100 kph both on the trial and the control runs. For the 4-mm and 6-mm markings, noise levels reduced with increasing spacing. The difference between the 50-mm spacing and the 450-mm spacing was about 5 dB(A). There appeared to be little difference among the spacings for the 8-mm markings, with the exception of the 250-mm spacing at 100 kph. This may be due to inconsistencies in the driving pattern. Rather surprisingly, there was little difference among the different heights of marking.

The same general features were found in the data for the four-wheel drive vehicle. In this case, the noise levels at 100 kph are clearly higher for the 4-mm and 6-mm markings, but not for the 8-mm markings. For the 4-mm and 6-mm markings, the 50-mm spacing produced noise that was approximately 7–8 dB(A) louder than the wider spacings. There was little difference between the 200-mm and 450-mm spacings. For the 8-mm markings, the results were more complex. At 100 kph, the lowest noise was generated by the 200-mm spacing, which was still 5 dB(A) louder than the control at that speed.

Only the 8-mm markings were tested with the truck, pilot runs having demonstrated that no noise increase could be detected by the experiment team. The 50-mm spacing generated noise approximately 4 dB(A) louder than the control with a laden vehicle travelling at 100 kph. With the unladen vehicle, noise was only 1 dB(A) greater. For the other spacings at 100 kph, the noise was approximately 2 dB(A) greater than controls. At both 80 kph and 60 kph, the maximum difference between the trials running on the edgelines and the controls was 1 dB(A) with the laden truck. Some of the runs with the unladen truck were in fact quieter than the controls.

Frequency analysis

Besides the overall noise level, the frequency characteristics of the noise are also important. If the noise generated shows concentration in one region of the audible spectrum, this will be heard as a hum of a certain pitch, quite distinct from the broad-band noise generated by the vehicle. This, rather than the increase in overall noise level, may be the critical factor in alerting the driver that the edgeline is being crossed.

The frequency analysis has been carried out in A-weighted decibels (dB(A)). This measure gives higher weightings to frequencies in the middle range of hearing and lower weightings to frequencies at the lower and higher ranges of the audible spectrum in a manner corresponding to the sensitivity of the human ear. Thus the analysis takes account of the response of the ear, and uses a measure which corresponds to the effects on the driver's hearing of the sounds generated by the different markings.

As the effect of the markings is to generate low-frequency noise, using dB(A) as a measure rather than unweighted decibels has the effect of reducing the distinct peaks which appear in the analysis. Nevertheless, the peaks are still easily discernible.

Frequency analyses were carried out on the noise generated by the Camry sedan. The results are summarized in Tables 14.1 and 14.2.

The data would seem to accord quite well with the predictions that could be made from a very simple application of first principles. The frequency with which the tyre hits a marking depends on both the speed of the vehicle and the spacing of the markings. If one striking event were the only source of additional noise, then the expected noise pattern would show a peak corresponding to the product of the vehicle speed in

Table 14.1 Comparison of observed peaks with predictions

Speed (kph)	Spacing (mm)	Predicted peak Hz	Approximate observed peaks Hz	
100	450	56	45	150
	200	11	90	300
	50	277	220	500
80	450	44	35	80
	200	89	70	
	50	220	320	400
60 (4 mm and 6 mm)	450	33	30	60
	200	67	60	110–170
	50	167	120	110–170
60 (8 mm)	450	33	42	90
	200	67	90	180
	50	167	220	420

Table 14.2 Difference in sound level between peak frequency and that same frequency on control run dB(A)

Speed (kph)	Spacing (mm)	4 mm	6 mm	8 mm
100	50	6.7	4.7	4.0
	200	−0.2	1.6	6.8
	450	−2.8	−4.5	1.8
80	50	−0.8	5.0	2.8
	200	2.3	6.2	5.2
	450	−6.1	−1.0	−3.1
60	50	4.0	−2.6	7.2
	200	4.8	2.9	5.8
	450	0.1	0.7	4.5

metres per second and the number of markings per metre. Difficulties in maintaining a uniform speed could be expected to result in minor deviations from the exact theoretical values.

It is also likely that this simple prediction would be complicated by a range of factors, such as a second impact when the wheel comes off the marking and contacts the pavement, the effects of the rear wheel running on the marking out of phase with the lead wheel, and the influence of the vehicle's suspension reducing impacts, particularly at higher speeds.

The predictions from this simple model for each combination of speed and spacing are shown in Table 14.1. Bearing in mind the difficulties in maintaining a steady vehicle speed and the difficulties of reading accurate values from logarithmic graphs, it is evident that for all combinations of speeds, spacings and heights of materials, a peak at or close to the predicted value was obtained, with the exception of the 8-mm material at 60 kph, where the peaks occur at close to twice the predicted value. This may be because the height of the material and the low speed generate a second major impact as the wheel contacts the pavement surface after crossing the marking.

In most cases, some second-order peaks are apparent. Most of these peaks are at about double the frequency of the first peaks, although in the case of the 450-mm and 200-mm spacings at 100 kph, they are at about three times the frequency of the first peak.

It is not certain how to characterize these peaks in a way which reflects their subjective loudness. On the one hand, reference to Table 14.2 shows that the energy levels of the peaks is, in general, only slightly higher than the A-weighted decibels for that combination of speed, spacing and height, with the exception of the 450-mm spacing, where in many cases the energy level at the peak is lower than the value for the A-weighted decibel. On the other hand, while taking the difference between the energy at the peak point in the spectrum and the energy at that point on the control runs shows clear increases over the control condition. The problem with this approach is that it takes no account of the possible masking effects of sound at different frequencies. Without further work, probably the most that can be said is that, judged by the second criterion, the 4-mm material and the 450-mm spacing nearly always generate peaks where the energy level is less than or close to the overall energy level, while the 6-mm and 8-mm materials at the 200-mm and 50-mm spacings nearly always generated peaks where the energy level was higher than for the dB(A) measure, and that there is no reliable pattern of differences between them.

Unfortunately, it is not possible to relate these characteristics of the sound directly to probability of detection under operational conditions. This could be done relatively simply by carrying out an experiment to determine the threshold for the peak noise against a background noise similar to that generated in the control runs, i.e. to determine the faintest noise at that frequency which can be detected against the background noise on 50 per cent of occasions. The noise generated by the different profiles can then conveniently be considered in terms of multiples of threshold, i.e. the number of times the signal is greater than threshold. Such work is beyond the scope of the present exercise, and unfortunately data which would enable an approximation to this have proved elusive.

14.4 Conclusion

The 6-mm and the 8-mm materials, at 200-mm or 50-mm spacings, generally gave an adequate signal for both the sedan and the four-wheel drive. This was reflected in both an increase in the overall noise level, and in pronounced peaks at particular frequencies in the audible range which could be related to the frequency with which the tyre hit the markings at different combinations of speed and spacing.

Despite the differences in the test method and the types of profile marking tested, the results of the present investigation were very similar to those of Soulage (1993) discussed in the introduction. In both cases, the internal noise levels inside saloon cars increased by 5 dB or over, up to 10 dB in Soulage's case, and up to about 8 dB in the present study. Soulage claims that his data showed that distinct peaks in the audible spectrum were generated with peaks proportional to the speed of the vehicle. In the present study, it has been shown that there is reasonably good agreement between these peaks and the frequencies which would be expected from the combination of speed and spacing.

The major point of difference concerns the noise generated by trucks. Although both studies agree that the overall increase in noise levels (i.e. the dB(A) measure) was very small, Soulage claimed that truck drivers were able to hear the change in pitch caused by running over the markings. In the present study, the experimental team reported that it was very difficult to hear any change in the noise. The increases in noise were especially small and unreliable with an unladen truck. One very high marking did produce a noticeable change in the overall noise level in the Soulage study, but this was twice the height of the highest marking available in the present study. Soulage rejected this very high marking on account of the very high level of environmental noise it generated.

It would therefore appear that none of the profile markings used in the present study could be considered to give an effective auditory cue to truck drivers, although they would still benefit from the improved optical guidance. It is possible that different proprietary brands of profile line marking laid by different processes may give different results, and may produce an auditory signal distinct enough to be heard in a truck. This possibility should be examined at the earliest opportunity.

References

Kryter, K.D., 1970, *The effects of noise on man*, New York: Academic Press.
Soulage, D., 1993, Effet sonore et vibratoire des marquages profiles, *Revue Générale des Routes et des Aerodromes*, **704**, Feb. 1993.

15 Detecting fatigued drivers with vehicle simulators

Anthony C. Stein

15.1 Background

The purpose of this section is to familiarize the reader with the general problems drivers face when fatigued; and to familiarize the reader with the development efforts leading to the research discussed in this chapter. A more complete synopsis is found in a companion paper (Stein, 1993), and a full background discussion is found in the Technical Report of this project (Stein et al., 1992).

Klein and his colleagues (Klein, Allen and Miller, 1980) synthesized the work of Mackie and Miller (1978). They found that as driving time increases, probability of accident involvement increases, while lane tracking ability decreases. Mackie and Miller also used the critical tracking task (CTT) (Jex, McDonnell and Phatak, 1966) as a secondary measure of truck driver performance. CTT data were gathered at various stopping points during the drive. The data trend found a decrement in CTT score as fatigue increases.

In general, they found that as the driver became more fatigued, performance decrements were observed. In the case of more severe fatigue, the decrement equaled alcohol impairment at levels of 100 mg.

During the mid-1980s, the state of Arizona conducted investigations of single-vehicle run-off-road commercial vehicle accidents and found the majority could be attributed to driver inattention caused by fatigue (Stein et al., 1989). The state faced a major obstacle in trying to prevent these accidents, however, there was no objective test which could be used by a patrol officer to determine if a driver was too fatigued to continue. Under current US law, an officer is allowed to place a driver 'out of service' if the officer determines the driver is either too ill or too fatigued to continue. This statute was rarely used because the officer was forced to accept the driver's word concerning the driver's hours-of-service. Thus, unless the driver's log indicated excess hours, or if the log was not filled out, the officer was forced to allow the driver to continue.

The desire of the Arizona Department of Public Safety (DPS) to reduce the occurrence of single-vehicle fatigue-related accidents led to the research discussed in this chapter. The first phase of the project involved development of the truck operator performance system (TOPS) hardware and software, preliminary testing to determine if the system was sensitive to driver fatigue, and initial feasibility testing to determine the operational viability of the system (Stein et al., 1989).

The results of this phase of the research led to a driving simulator based system which proved sensitive to fatigue, and received high ratings for realism and face validity by randomly selected truck drivers.

The second phase of the contract included gathering performance and fatigue data from on-the-road drivers passing through the state's ports-of-entry, and developing pass/fail algorithms from the available databases (Stein et al., 1990). This phase also proved successful, with the algorithm having the potential of identifying approximately 50 percent of the drivers identified as impaired, while only failing five percent of the non-impaired group. These statistical test properties, while not ideal, were acceptable. It should be understood that performance-based testing will never be 'perfect' in its discrimination. Because the state was interested in detaining the drivers most likely to be involved in an accident, and these were expected to be the most impaired, a 50 percent detection rate will most likely pick up the target group. Also, since the test was designed with a selective enforcement model in mind, and not random testing, the five percent false-positive rate only comes into play when a driver suspected of being overly fatigued is tested. Since the five percent figure comes from a random group of drivers, the actual false-positive rate is expected to be lower.

The final phase of the project involved controlled tests to determine the correlation between poor performance on the TOPS and actual driving simulator performance. Physiological data were gathered from a subset of the test population to see if the performance degradation correlated with indicators of reduced alertness. This phase of the project is the subject of this chapter.

15.2 TOPS Validation Experiments

Purpose

The purpose of the validation experiments was to determine whether impairment measured by the TOPS was consistent with general driving impairment, and with loss of alertness. To measure general driving behaviour the TOPS facility was converted, through the flexibility of the Systems Technology, Inc. (STI) low-cost simulator, to a fully interactive driving simulator (described below). Loss of alertness was examined in selected subjects with physiological measures. Both of these testing methods will be discussed fully in the following sub-sections.

Methods

Basic experimental approach
The basic experimental approach of this study involved studying the effects of extended driving time on driving behaviour in a fully interactive fixed-base driving simulator, and on a performance test developed to detect impairment due to fatigue (the TOPS).

Cumulative workload was established by having the subjects drive for specified time periods, with specified break and TOPS testing times. Each subject came to the test site for four consecutive days of testing. Day 1 through day 3 each required 10 hours of participation, and day 4 required 20 hours of participation. Physiological measurements were gathered on selected subjects to correlate driving and TOPS behaviour with levels of alertness (discussed below).

Experimental design
Twenty subjects were tested using a repeated measures design. Data were gathered once each hour using both an interactive driving simulator and TOPS. Six of the subjects also provided physiological data to determine their level of alertness. Each subject reported for four consecutive days, with the first three days lasting 10 hours, and the final day lasting 20 hours.

Twenty volunteers were selected from employees of a major trucking firm. Subjects were required to take vacation time, or time off without pay, to participate in the study. Criteria for selection included:

- current licensure as a truck driver;
- a current Department of Transportation medical card; and
- abstinence from alcohol and drugs (other than required prescription drugs) for the duration of the study.

Because of the requirement for a valid commercial drivers licence, the minimum age for participation was 21. The resulting population included a broad spectrum of drivers. The population had the following characteristics.

- 18 males
- 2 females
- age range: 23–62 years (mean = 42.9, sd = 9.5, median = 44)
- driving experience: 2–31 years (mean = 12.0, sd = 9.4, median = 7)

Apparatus

The driving simulator and the TOPS shared the same facility. Software flexibility resulted in the ability of the system to serve this double duty. The driver's area included a mock-up of an actual Class 8 truck, with the dashboard, steering wheel, seat, and brake and accelerator pedals taken from an actual vehicle. The area was built so the simulator display was placed on the top of the dashboard. Thus, the roadway was presented in the driver's line of sight, and appeared similar to an actual roadway with an horizon scene and side task boxes on either edge of the display (Figure 15.1).

The experimenter's area included the computer, a keyboard and an experimenter's monitor. From this location, the experimenter was able to select the system mode (i.e. simulator or TOPS), and begin the run. The system automatically recorded the data for later analysis. In the case of the TOPS, 'pass' and 'fail' scores were also available.

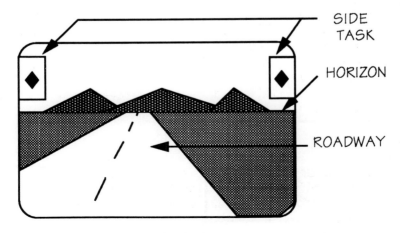

Figure 15.1 The Simulator Roadway Scene.

Driving simulator tasks and measurements

The driving simulator tasks were based on tasks used in previous simulator research. These tasks have been found to be sensitive to impaired driving performance, and are similar to tasks which would be encountered during normal driving. The measures selected for analysis, discussed later in this chapter, represent both basic traffic safety indicators, and behaviours which have been shown to be indicators of impaired driving performance which could result in accident involvement (Allen and Stein, in press; Stein 1987; Stein and Allen, 1986; 1987; Stein et al. in press).

The tasks were presented as a scenario, which simulated a drive between two points, and required the driver to negotiate curves, avoid obstacles, and maintain speed. The driver was also required to respond, periodically, to a side task which was displayed at the edge of the monitor. The driver's impression was one of driving on a rural roadway, with mountains in the background. Periodically he or she needed to perform some control action (e.g. steer, brake, etc.) to avoid having an accident or respond to a sign.

A traffic officer was placed on the course in various locations, and if the driver exceeded the speed limit by 5 kph (3 mph) when the officer was present a ticket was given. When the driver received a ticket, a siren sound was played indicating a ticket had been received. The driver was only required to slow down to the speed limit. If the driver had an accident (by running off the road), or was involved in a collision (striking another vehicle), the display showed a cracked windshield, a loud crash sound was played, and the screen froze for a short period of time.

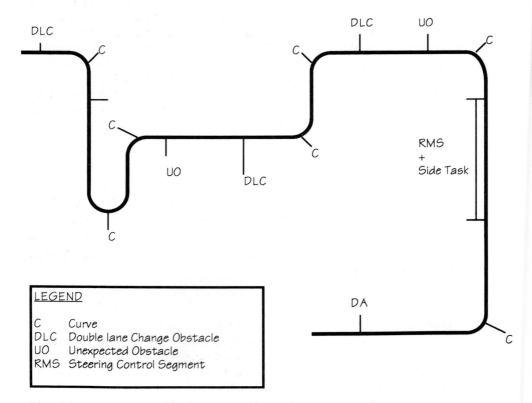

Figure 15.2 A schematic of a driving scenario.

A typical driving scenario consisted of a road approximately 67.6 km in length (40 miles) during which various events were encountered. A total of nine different scenarios were created. Each had equal numbers of the various events, but had the event order changed. Since each driver drove 45 times, each saw a given scenario only five times. Order presentation was randomized, so the probability that a given scenario was memorized was slight. The events occurred at various locations along the roadway, thus event presentation timing was proportional to vehicle speed. A typical scenario segment is shown in Figure 15.2, and a discussion of the scenario events follows.

Steering control with divided attention
Degraded steering control was detected through the use of the sub-critical tracking task (Stein *et al.*, 1990) which was introduced into the driving task by adding it into the vehicle's lateral (steering) vehicle equations of motion. When added to the simulation in this manner, it appeared to the driver that the vehicle was being disturbed with random wind gusts. The sub-critical tracking task was introduced into the lateral dynamics at a level which required the driver to maintain fairly constant vigilance to keep the vehicle travelling in the correct location in the lane ($\lambda = 0.15$).

The side task was added to provide an additional workload demand on the driver, and to add a task which required the concurrent performance of steering control and visual search. In the upper right and left portions of the screen were indicators which contained various symbols (Figure 15.1). During the majority of the time the symbol was a diamond. Periodically (approximately six times per minute) the symbol in one of the task boxes changed to either a triangle or horn. If the symbol changed to a triangle, the correct response was moving the turn indicator lever in the direction of the triangle. If the symbol changed to a horn, the correct response was depressing the horn button in the centre of the steering wheel. The symbol returned to the diamond when any indicator was activated, or at the end of two seconds if no indicator was activated. Drivers were instructed to respond appropriately to the task as quickly and accurately as possible. They were also informed of the time/accuracy trade-off in the reward–penalty structure (discussed later), which emphasized accuracy.

Obstacles
Two types of obstacles were presented to the driver:

- an expected obstacle which required the driver to manoeuvre between three vehicles, called the **double-lane-change (DLC) obstacle** (Figure 15.3(a)); and
- a single vehicle in the driver's lane of travel which was not visible until it was close enough to require immediate action to avoid a collision, called the **unexpected obstacle** (UO) (Figure 15.3(b)).

Curves
The curve event (C) required the driver to control both speed and vehicle position. The simulator allowed the experimenter to determine both the radius of curvature and the length of the curve. Thus curves could be gentle, and require only steering control; or they could be sharp, and require the driver to slow to a speed which would allow the curve to be negotiated without losing tyre traction and running off the road. The curves were defined prior to beginning the experiment, and equal numbers of each curve were

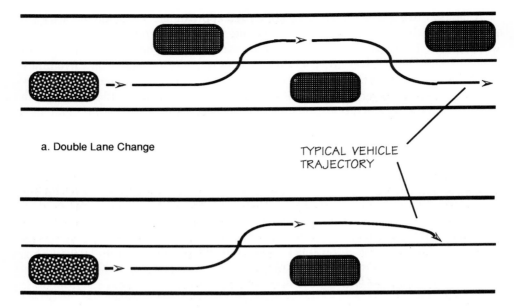

Figure 15.3   Ground plane representation of the obstacle tasks.

included in each of the nine scenarios. This task is sensitive to driver fatigue which commonly results in the driver running off the road (Sweedler *et al.*, 1990).

Simulator data collection

As part of the normal operation of the simulator, data were collected at 20 Hz. The data collected include eight continuous and eight discrete variables. The continuous variables were:

- steering activity
- lane position deviation
- heading error
- curvature error
- vehicle speed
- vehicle acceleration
- throttle activity
- response time to the side task.

The discrete variables were:

- simulator accidents (running off the road)
- simulator collisions (striking another vehicle)
- speed exceedances (the number of times the speed limit is exceeded by over 5 kph)

- speeding tickets (the number of speed exceedances when the traffic officer is present)
- run completion time
- correct responses to the side task
- incorrect responses to the side task
- missed responses to the side task.

TOPS tasks and measurements

The discussion here focuses on the differences between TOPS and the simulator. Instead of being presented with a series of curves, obstacles, and other tasks, the driver was placed on a straight road with no other traffic. Steering and speed control tasks were implemented by imbedding the sub-critical tracking task in both the lateral (steering) and longitudinal (speed) vehicle dynamics continuously, as opposed to the simulator where the tracking task was presented as a specific scenario element. The side task was also used, but less frequently (approximately two-and-a-half times per minute). The driver's impression was one of driving down a straight road, with no other traffic. Wind gusts are felt, and occasionally a side task symbol needed attention.

The traffic officer was *not* placed on the highway, so if the driver exceeded the speed limit there was no penalty. Also, if the driver had an accident, the system reset the vehicle back in the centre of the roadway, and no crash sound was heard or visual information concerning a crash given. This was to reduce the possibility of arousing the driver.

The test lasted eight minutes, and upon completion of the TOPS test, the data were analysed on-line. The results of the analysis were compared with previously established pass levels, and the system determined whether the driver passed or failed the test.

Physiological measures

As discussed above, six of the 20 subjects were fitted with physiological measurement devices during the experiments – an eight-channel (plus time) Medilog 9000–II ambulatory recording device. These subjects were also asked to wear an actigraph for the week before their testing. The actigraph is a wristwatch-sized activity monitor, and was used to determine that the subjects had 'normal' sleep–wake cycles. These subjects were also asked to complete a sleep log from the time they received the actigraph until they completed the experiment. This log was used to record periods of sleep and activity.

From the beginning of the third day through to the end of the experiment, the following data were collected continuously:

- electro-encephalogram – EEG,
- electro-oculogram – EOG, and
- muscle tone – EMG.

15.3 Experimental procedures

Training

Prior to the experiments, all subjects were given basic training on the simulator tasks, and on TOPS. Driving the simulator was similar to driving a new piece of equipment. The driver knew basically what to do, but was not familiar with the vehicle's subtleties and handling characteristics which may be required to negotiate the driving scenario. Also, there were tasks which the driver had not performed before participating in the test (i.e. the side task). During the training session, the subject was given basic familiarization on the system, and shown each of the events which would be part of the driving scenario. The subjects were allowed to practise driving the simulator through each event until they felt comfortable with their performance. This training was not intended to allow the driver to reach asymptotic behaviour, but rather to have the driver capable of completing the driving scenarios. Typically, drivers learned the various simulator tasks in about an hour (range 45–90 minutes, median 60 minutes).

We expected to have the drivers become more proficient in their driving as time progressed. This learning which occurred over time had the effect of making the data analysis conservative. For example, run time stabilized over time, making the detection of statistically significant differences more difficult. Because performance improvement was expected, if decrements were observed they were more likely to actually be due to extended driving hours rather than chance.

The TOPS training used the automated training regimen that is part of the system (Stein *et al.*, 1990). Basically, the system took the driver through each phase of the TOPS test, and determined that the driver met a minimum performance level before allowing the TOPS test to be administered. The training regimen took 5–10 minutes to complete. If the driver was unable to pass the minimum performance level after ten minutes, the system operator was notified and remedial training was given. In these experiments, all subjects completed the TOPS training using the automated system.

Reward–penalty structure

A reward–penalty structure was included as part of the simulator runs to motivate 'normal' driving behavior (Cook *et al.*, 1981). If this type of artificial inducement is not incorporated into the experiment, drivers are likely to behave in a way which would reflect their behaviour in an arcade game rather than on the road. This is because there were no real-world penalties for poor performance.

The reward–penalty structure amounts were selected to induce certain behaviours. First, we needed to ensure that if a subject began the study he or she would finish unless there was a compelling reason to quit. Second, we wanted the drivers to be able to earn about the same income which would be earned if they were driving rather than participating in the study. This led to participation pay. The drivers would receive this amount of money regardless of their performance. Next, we wanted to ensure the drivers were motivated to drive as they would in the real world. This led to the creation of rewards and penalties for various driving behaviours. If the driver performed appropriately, it was possible to more than double the participation fee.

Rewards were given for:

- completing the scenario, simulating the real-world motivation of arriving at a destination;
- quickly responding to the side task, simulating being aware of potential dangerous situations; and
- beating a reference time which was accomplished by driving at about the speed limit, simulating the real-world motivation of driving 'with the flow of traffic.'

Penalties were assessed for:

- going slower than the reference time, simulating being late for a pick-up or delivery;
- having an accident by running off the road;
- having a collision by striking another vehicle;
- missing the side task; and
- receiving a speeding ticket.

To help ensure participation for all of the experimental sessions, monies were paid at the end of the subject's experimental participation, and bonus monies could only be collected if the experiment was completed. The components of the reward-penalty structure are given in Table 15.1. The reference time was established to allow the drivers to 'beat' it by 2–3 minutes if they drove at about the speed limit. The penalty may seem low for accidents and collisions in relation to the real-world costs, but it was chosen for several reasons. If we were to appropriately simulate the real-world contingency for having an accident or collision, we would need to eliminate the driver from the study at the first collision. This would have been disastrous for two reasons:

1. we had both time constraints on completing the study and there was no waiting list of potential subjects; and
2. we would have lost valuable data.

We treat accidents and collisions in the simulator as a dependent variable. They provide insight into the behaviours which lead to real-world situations.

Experiments

Each subject was tested over a period of four consecutive days. Day 1 through to Day 3 lasted just over 10 hours, and included 10 TOPS trials, nine 45-minute simulator runs, and periodic rest breaks. Day 4 lasted 20 hours, and consisted of two 10-hour schedules back-to-back. The experimental day began around 7 a.m. with the 10-hour day ending around 5 p.m., and the 20-hour day ending around 3 a.m.

Table 15.1 Reward–penalty components

Item	Reward	Penalty ($)
Participation monies	Daily pay (Days 1–3)	50
	Daily pay (Day 4)	100
Run-related monies	Run completion bonus	3/scenario
	Time saved bonus	1/min
	Time lost penalty	1/min
	Accident and collision penalty	2 each
	Ticket penalty	1 each
	Side task correct response	0.01 to 0.10
	Side task incorrect response	0.10

15.4 Results

Data analysis

A commercial statistical analysis program (Statistica Ver. 3F, StatSoft Corp, 1991) was used to perform both multivariate analysis of variance (MANOVA) and stepwise linear regression analyses of the appropriate data sets. There were actually three data sets:

- the TOPS data, which were analysed to determine the current algorithm's ability to detect fatigued drivers;
- the simulator data, which were analysed to determine decrements in driving performance as a function of fatigue; and
- the physiological data which were analysed to determine lapses in alertness as a function of extended driving.

The objects of the analyses were to:

- determine whether simulator driving performance was degraded as a function of extended driving; and
- determine the correlations between degraded TOPS performance and decreased driving capability, and whether these are a function of extended driving.

The physiological measures were included to determine if performance was correlated with alertness.

TOPS data

The algorithms developed during the second phase of the project relied on either a before/after categorization, a DPS officer's subjective assessment of a driver's state of fatigue, or on an experimental design condition. They were not based on how long the driver had been working, or the workload. Thus, the previous data analyses, while appearing to be adequate in detecting impaired drivers, could have been found to be invalid if the assumptions grouping the data were in error. We did not believe this to be the case, and the analyses performed during this phase of the project proved our assumptions true.

The MANOVA found that the increase in TOPS failure, as a function of extended driving, on Day 4 was statistically significant ($p = 0.002$, $F = 7.196$). The results of the Phase 2 work suggest that approximately 50 percent of fatigue-impaired drivers should be detected and that the unimpaired failure rate should be approximately five percent. With only 20 subjects, determination of the five percent failure rate is not possible because each failure accounts for a high percentage. In this study, the failure rate for the first 10 hours of Day 4 averaged 14 percent ($\sigma = 5.4\%$; 3 failures ± 1). Also, the detection rates did not reach the predicted 50 percent level. Since it is probable that not all of the drivers were impaired due to fatigue, even at the end of the 20-hour day, a failure rate below the 50 percent level is not unexpected.

Simulator performance

The simulator performance data were analysed using multivariate analysis of variance techniques (MANOVA). The data have been divided into two basic measures of driver

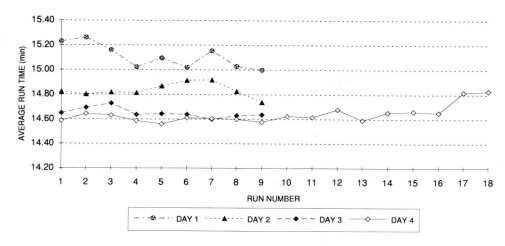

Figure 15.4 Run completion time.

performance for discussion purposes: overall performance and actual performance. Overall performance, indicates how the driver's performance would be observed in the outside world. Data in this area include:

- accidents (running off the road)
- collisions (striking another vehicle)
- speed exceedances
- tickets (speed exceedances when the traffic officer was present)
- scenario completion time (an indication of the persons ability to drive with the flow of traffic)

In general, these data show the overall trends observed throughout the experiment. During the first three days, learning was occurring, resulting in improved performance. This is particularly evident in Figure 15.4 where driving times decreased as a function of test day ($p < 0.001, F = 226.72$).

For the data shown in Figure 15.5, at about Run 13 performance begins a consistent deterioration, rather than moving about the mean ($p < 0.017, F = 4.30$). This is an important finding in the study, because this data element represents behaviours which contribute to fatigue-caused truck accidents (Sweedler et al., 1990). This is in contrast to simulator collisions where no apparent differences are seen ($p = 0.320$).

As with simulator accidents, the other overall performance data elements identified statistically significant deterioration in driver performance throughout Day 4. The results were:

- speed exceedances – $p = 0.019, F = 4.20$
- tickets – $p = 0.025, F = 3.88$
- run time – $p = 0.008, F = 5.26$

The second type of driver performance data consists of those variables which describe the driver's actual performance. While these data, for the most part, do not indicate

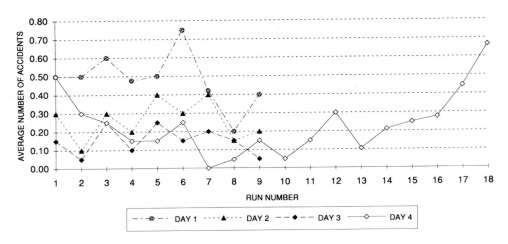

Figure 15.5 Simulator accidents.

driver failures resulting in observable actions, they are indicative of driving errors which can result in accidents.

The data were divided into several time epochs and analysed using multivariate analysis of variance techniques to determine the statistical significance of the differences. Subjects were treated as a random variable. Data from the curves were analysed first, but due to the large between-subject variance, few statistically reliable results were observed. The data on the straight segments were analysed next. Here, the between-subject differences were small, as drivers approached the tasks in a similar manner. For this chapter only limited analyses of key variables were selected for discussion. These variables represent each of the primary tasks the driver is required to perform while driving the simulator:

- maintain appropriate lane position
- maintain appropriate speed
- respond to side-task stimulus.

In Figure 15.6 we see the plot for the standard deviation of lane position ($p < 0.001$, $F = 14.960$). This is probably the most important indicator of fatigue-based impairment (Klein et al., 1980), and as this value begins to rise above one foot, there is a geometric increase in the probability of exceeding lane edge boundaries (Allen and Stein, in press). As with the TOPS performance, the drivers' performance remained fairly constant until Run 13, when consistent deterioration in performance began. In this case, by Run 17, SDLP had increased into the critical region.

Another lane position variable, standard deviation of heading error, indicates the variability of the error in the drivers' tracking performance. Again, performance was fairly stable until Run 13, when the variability began to increase consistently ($p < 0.001$, $F = 19.308$).

Another indicator of impaired driving performance is the variability in a driver's speed. As with the other data, performance remained stable, around 0.5 mph, until Run 13. At this point, the driver began to lose the ability to maintain speed, and the variability began an upward climb ($p < 0.001$, $F = 32.673$).

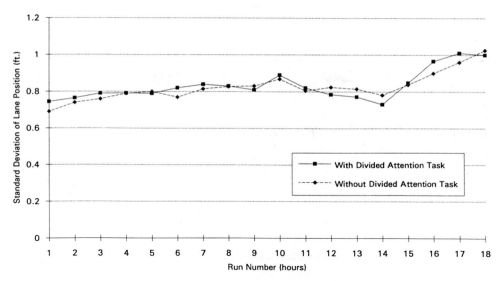

Figure 15.6 Standard deviation of lane position (SDLP) during the tracking task.

Finally, similar performance decrements were found in the drivers' reaction time to the side task. Typically drivers respond to these stimuli in about 1.5 seconds, and in this study we found the same results for unimpaired drivers. Again, however, we found that, at about Run 13, the drivers began to exhibit degraded performance. In this case, reaction times increased 33 percent to approximately 2 seconds ($p < 0.001, F = 21.273$).

Physiological measures

The physiological data were analysed using automated methods described fully in the technical report (Stein *et al.*, 1992). Basically, the data were automatically scanned using software which detected differences in subject arousal state. The algorithm used a non-uniform epoch length, with a new epoch beginning whenever a state change was detected. The data are categorized using artificial intelligence (AI) algorithms which utilize a decision tree strategy.

Data analysis included scanning of the nocturnal sleep portions of the recordings to determine whether any sleep abnormalities were present. No such abnormalities were found, and the subjects appeared 'normal' for the purposes of these tests.

The analysis identified the percentage of time the subject spent in each stage of alertness. Stages one to five indicate the subject was sleeping, and in these experiments there was only one occurrence. Stages six and seven indicate drowsiness, and stages above seven indicate different levels of wakefulness. In general, the subjects tended to be awake throughout the testing, however an increase in stage seven drowsiness was observed in the second half of Day 4.

Unfortunately both the physiological and TOPS/simulator data analyses were conducted concurrently, so the sub-contractor responsible for the analysis of the physiological data was unaware of the apparent onset of impaired driving behavior beginning at Run 13. Thus their data analysis, which split Day 4 into two equal periods is not as robust as it might be. Nevertheless, the trend of increased drowsiness in the second half of Day 4 is consistent with the subject's reported level of fatigue, the TOPS data and the simulator performance data.

Comparison of TOPS and simulator performance

Multiple regression techniques were used to compare the drivers' performance on the simulator and TOPS tests. Both standard and forward stepwise regressions were conducted comparing TOPS failure rates with simulator performance. Both SDLP and standard deviation of speed were found to correlate positively with TOPS failure (i.e. as performance on these variables deteriorates TOPS failure increases). For SDLP the correlation was 0.623 ($p = 0.006$), and for the standard deviation of speed the correlation was 0.883 ($p = 0.003$).

These two variables are indicators of potential driving problems. As discussed above, increases in SDLP correlate with the probability of exceeding lane edge boundaries, which can result in an accident. Increases in speed variability increase the probability that the driver will enter a speed-limited curve at too high a speed, resulting in a run-off-road accident.

15.5 Discussion and concluding remarks

The purpose of this study was to determine the correlation of TOPS performance with driving performance and physiological data under controlled conditions. If this was successful, then there is good reason to believe that TOPS is a useful tool for removing fatigue-impaired drivers from the transportation system.

Because data were analysed using mean scores for all subjects, it was not possible to correlate directly a given individual's performance on TOPS vs the simulator. As a group, however, the drivers' TOPS performances correlated very well with the simulator driving performance (i.e. poorer TOPS performance correlated with poorer simulator performance).

In the experiments, TOPS performance for the first 10 hours on Day 4 remained fairly constant. After that time it began to deteriorate. While the early failure percentages are higher than the predicted five percent, the small number of subjects means that a single failure had a large effect on the unimpaired failure rate. It is also possible that the drivers failing the test were actually impaired, despite the assumption to the contrary.

The simulator measures three basic tasks involved in driving:

- the steering control capability of the driver;
- the speed control capability of the driver; and
- the divided attention capability of the driver.

Steering control is important because it is necessary to maintain the vehicle in the proper lane of travel, and speed control is important because driving at a speed vastly different than the flow of traffic creates problems which may result in an accident. As driving is considered a divided attention task, the simulator measures the driver's capability to attend to more than one task at a time. Degradation in any, or all, of these areas may result in an accident.

The simulator data indicate increasing impairment from Run 13 onward on Day 4. This impairment is evident in one of the key traffic safety variables, accidents, where the average number of simulator accidents increased from about 0.15 accidents per run for the first 12 hours to 0.68 accidents per run for hour 18, over a fourfold increase ($p = 0.017, F = 4.306$). In the simulator, accidents are defined as running off the road. This can occur because the driver loses awareness and drifts off the road on a straight, or because he or she fails to slow sufficiently for a curve and loses traction. Both of these cases are indicative of a loss of alertness, and are the type of accident frequently associated with fatigue-impaired driving (Sweedler et al., 1990).

A variable considered an important indicator of fatigue-based impairment is SDLP (Klein et al., 1980). This variable is better known as 'weaving', and when the weaving behaviour increases beyond about one-foot, there is a geometric increase in the probability of exceeding lane edge boundaries (Allen and Stein, in press). Data from the simulator runs indicated an increase in SDLP beginning at Run 13 and by Run 17 SDLP had increased into the critical region ($p < 0.001, F = 14.960$).

The steering control decrements discussed above are reflected in both the driver's ability to maintain speed control, and to attend to divided attention tasks. In the former, the driver's variability in maintaining speed control increases almost 150 per cent ($p < 0.001, F = 32.673$); and in the latter, the driver's response to a secondary task takes about one third longer ($p < 0.001, F = 21.273$). Both of these performance decrements begin at the thirteenth hour, and increase steadily.

The physiological measures are not as well defined in terms of identifying impairment, but they do indicate increases in less wakeful states. In this study, the Day 4 data were divided into equal parts, and a slight increase was observed in drowsiness in the second half of the day.

The purpose of these experiments, stated above, was to determine the correlation of TOPS data with simulator performance and physiological measures of decreased alertness. As discussed, both TOPS performance and simulator driving performance deteriorated as a function of driving time, particularly after Run 13 on Day 4. Here there were positive correlations between TOPS failure and SDLP (correlation = 0.623, $p = 0.006$), and between TOPS failure and ability to maintain speed control (correlation = 0.883, $p = 0.002$).

While the physiological data were not as robust as the TOPS or simulator data, they still showed trends in the same direction. From these findings, it is apparent that the TOPS measures functions of driver performance which correlate with both actual run-off-road accidents (in the simulator), and with measures of impaired driving; and that these performance decrements are a function of extended driving periods.

References

Allen, R.W. and Stein, A.C., in press, The driving task, driver performance models and measurements, in O'Hanlon, J.F. and de Gier, J.J. (Eds) *Drugs and Driving–II*. Philadelphia: Taylor & Francis.

Cook, M.L., Allen, R.W. and Stein, A.C., 1981, Using rewards and penalties to obtain desired subject performance, *Proceedings of the Seventeenth Annual NASA – University Conference on Manual Control*, Pasadena, CA: Jet Propulsion Laboratory.

Jex, H.R., McDonnell, J.D. and Phatak, A.V., 1966, *A 'Critical' Tracking Task for Man–Machine Research Related to the Operator's Effective Delay Time: Part 1: Theory and Experiments With a First Order Divergent Controlled Element*, Hawthorne, CA: Systems Technology, Inc. NASA CR–616.

Klein, R.H., Allen, R.W. and Miller, J.C., 1980, *Relationship Between Truck Ride Quality and Safety of Operations: Methodology Development*, Hawthorne, CA: Systems Technology, Inc., NTIS: DOT HS 805 494.

Mackie, R.R. and Miller, J.C., 1978, *Effects of Hours of Service, Regularity of Schedules, and Cargo Loading on Truck and Bus Driver Fatigue*, Goleta, CA: Human Factors Research, Inc., Report 1765–F.

Stein, A.C., 1987, A simulator study of the effects of alcohol and marihuana on driving behavior, in Noordzij, P.C. and Roszbach, R. (Eds) *Alcohol, Drugs and Traffic Safety – T86*, New York: Excerpta Medica.

Stein, A.C., 1993, Vehicle simulation and fatigue, in the *Proceedings of the Chartered Institute of Transport in Australia National Conference*, Perth Western Australia: Chartered Institute of Transport.

Stein, A.C. and Allen, R.W., 1986, The use of in-vehicle detectors to reduce impaired driving trips, in Viano, D.C. (Ed.) *Alcohol, Accidents, and Injuries*, pp. 123–30, Warrendale, PA: Society of Automotive Engineers.

Stein, A.C. and Allen, R.W., 1987, The effects of alcohol on driver decision making and risk taking, in Noordzij, P.C. and Roszbach, R. (Eds) *Alcohol, Drugs and Traffic Safety – T86*, New York: Excerpta Medica.

Stein, A.C., Allen, R.W. and Parseghian, Z., 1990, *High Risk Driver Project: Development of Methods for Testing Truck Driver Fatigue*, 2 vols, Hawthorne, CA: Systems Technology, Inc., TR–2387–1.

Stein, A.C., Allen, R.W. and Parseghian, Z., in press, *The Effects of Low Dosages of Alcohol on Truck Driver Control and Simulator Performance*, Hawthorne, CA: Systems Technology, Inc.

Stein, A.C., Aponso, B.L., Rosenthal, T.J. and Allen, R.W., 1989, *High Risk Driver Project: A Study of Truck Driver Fatigue*, Hawthorne, CA: Systems Technology, Inc, TR–2371–1.

Stein, A.C., Parseghian, A.C., Allen, R.W. and Miller, J.C., 1992, *High Risk Driver Project: Theory, Development and Validation of the Truck Operator Proficiency System (TOPS)*, 4 Vols, Hawthorne, CA: Systems Technology, Inc.

Sweedler, B.M., Quinlan, K. and Brenner, M., 1990, Fatigue and drug interaction in fatal to the driver truck crashes, in Petrocelli, E. (Ed.) *Proceedings of the 34th Annual Conference of the American Association for Automotive Medicine*, pp. 491–504, Arlington, IL: American Association for Automotive Medicine.

SECTION FOUR:
Empirical analyses of the impact of fatigue

16 The road transport industry: the company driver's perspective

Graham Derham and Colin Harwood

16.1 Driving to the north of Western Australia: Graham Derham

We arrive at the yard at 9 a.m. on Friday mornings for our two-up scheduled run to Kununurra, in the north of Western Australia, 3300 km from Perth. First we check out the dock to see what freezer/chiller freight we have and work out how many pallet spaces we will need in the freezer section of the trailer. Dry goods are loaded in the middle section. Then we put bulk heads up and load the chiller goods. We finish loading two trailers at about 2.30 p.m. and one is driven by one of the two drivers to the road train assembly area at West Swan, outside Perth, ready to be picked up for the trip north. The third trailer is loaded by other employees. It is taken through to Carnarvon, in a two-trailer combination. At Carnarvon, at the Shell truck stop, we hook up the third trailer to become a triple-trailer road train.

At the time of writing this, we are loading freezer/chiller goods and general freight on the first trailer, with a third flat top trailer being loaded with mining equipment for Cadgebup. While this is happening, the co-driver starts work on the second trailer. The first driver returns and helps his mate finish loading the second trailer and then we pick up our paperwork. We drive to the assembly area with the second trailer, hook up the first trailer and we are off, leaving the assembly area at 8 p.m.

The main destinations are Fitzroy, Halls Creek and Kununurra, driving the coastal highway. Our first stop on the way is Dongara, at the Shell truck stop, for a cup of coffee and to check the load. We stop for about half an hour. We work a four-hour rotation, stopping every four hours to swap the driving over. Our next stop is at Billabong roadhouse to check the tyres, the fridge on the freezer/chiller trailer and swap over drivers. The next stop is Carnarvon, 905 km from Perth, where we hook up the third trailer, often stopping for only ten minutes. The prime mover we are driving is the latest Mack, 525 hp. We do not take on any fuel until we reach Hedland. We carry 2000 litres of fuel.

We stop again at Barrdale, which again is another four hours on. Then it is another four hours to Karratha, about 1600 km from Perth, where we stop and change over. We do not eat at roadhouses; we take our food with us from home and have something to eat along the way. Our first sit-down meal is usually at Halls Creek on Sunday night. The next stop is Hedland, then Sandfire, the Broome turnoff and onto Fitzroy, our first unloading destination. On the way home, our routine is a bit different. Because we spend most of Sunday doing the physical work of loading for the trip home, for the first 12 hours or so we only drive two-hour stints between changeovers. After that we return to our four-hours shifts at driving. There is no problem sleeping in the sleeper cab. It is absolutely beautiful – best part of the truck. Our new truck is all on air suspension. The cab, the seats and the sleeper bunk is on air suspension. The sleeper cab is closed off, separately air-conditioned, just like a double bed.

Of course, it is important you have faith in your co-driver. You have to get along with him. After all, you see more of him than you do of your wife. In fact, two of my sons both work with me as co-drivers. The younger son has been with me on the same run for ten months at Key Transport. He started as a loader, worked as a forklift driver, before joining me in the truck. At Key Transport, drivers work their way up through the ranks until a driving spot becomes available, unless, of course, they join the company already having a lot of experience. Then they are given the chance to go straight onto the road. You can keep driving as long as you want, until you feel you have had enough. Management does not come to you and say because you are 45 or 50 or whatever, it is time to quit. That is up to you – to know when you've had enough. I have been driving trucks all my life. I bought my first truck at the age of 21, a small tipper, a JS Bedford, with a 300-cubic-inch petrol motor. In 1974, I began working with Bell Freight as a subcontractor, now I am a company driver. It is much easier being a company driver. You do not have the financial worries. Of course, with a company truck, you have to look after it, as if it is yours.

With driving, the number one requirement is experience. But everyone still needs to be taught. You can not gain experience unless someone is prepared to teach you – and you have to be taught the right way and the way the company wants to work. When you are a truck driver, apart from being a driver, you have to be a mechanic, an electrician, a tyre fitter, truck washer, all sorts of things. You have to be versatile.

My reaction to fatigue is, as far as I can see, it is tiredness. You are the only person who can tell when you have had enough. Driver fatigue can cause accidents but we can all have accidents without being tired. Someone can jump out of bed at 8 a.m. and by 9 a.m. he has smashed his car up on the way to work, without getting behind the wheel of a truck. When you are a long-distance driver, away from home, you have to pace yourself. It is the driver's responsibility to have the load at the destination within a reasonable time. There is no pressure from the company. The driver knows the schedule. We work our own hours. We self-regulate. A lot of drivers can not work two-ups but I find working two-up is the best way to go. You are only doing half the work, driving half the distance and there are two of you to deal with any problems.

My current routine, on the Kununurra schedule run, is we leave Friday and return home Wednesday. Thursday is my day off. When I arrive home I sit down, talk to my wife, watch TV, wind down, go to bed around midnight and sleep right through. In a month, you do two long trips and one short trip, an overnight run. So in one week of the month you have three days at home and have one week off per month too.

Alcohol and drugs are not a problem in the industry in Western Australia, as far as I know. You can not drink and drive any more. Those days are gone. Any driver who does not do the right thing does not last. He has gone straight away. The company has a lot of money tied up in the vehicles. We earn good money. We know if we look after our job for 52 weeks of the year we are going to have something. It is a profession.

I have never been involved in any accidents. The worst thing that has happened, two years ago, is when we did a wheel bearing coming out of Karratha one evening. I looked out of the rear vision mirror and thought that light's a bit bright. By the time we pulled up, the trailer was ablaze. An inside bearing had given out, and became so hot it burnt the tyres and the side of the freezer/chiller trailer. We threw water and sand all over it, put the fire out and drove back to Karratha. We saved the rockmelons.

On the family side, without a good woman at home, it just can not work. It is a good lifestyle but you need a good woman to back you.

Driving to Argyle Diamond Mine: Colin Harwood

I drive to many destinations in Western Australia. One of the longer routes is to Argyle Diamond Mine, near Kununurra in the north of Western Australia, about 3300 km from Perth. A working day for a two-up schedule to Argyle Diamond Mine loaded with mining equipment, starts at 10 a.m. on a Friday. We work through, apart from lunch, to between 6 and 7 p.m. Then we set off, with each driver doing three to four hours in rotation. For the Argyle trip, we travel the inland road, (Great Northern Highway) to Port Hedland, about 1500 km from Perth, then through Fitzroy to Hall's Creek and on to Argyle, arriving on Sunday at lunchtime. For the return trip from Argyle to Perth we depart around 11 p.m. on the Sunday, arriving back in Perth on Tuesday evening between 8 and 9.00 p.m.

The first driver changeover usually happens at Wubin, 272 kms form Perth; then we stop at a parking bay 40 km south of Mount Magnet, another 257 km further on. The third changeover is at Meekatharra which is 765 km from Perth. We top up with fuel at Meekatharra. The next changeover is usually 295 km further north. Mungina Gorge is the next changeover and again at Port Hedland to refuel. And so on to Broome, 2100 km from Perth, to Hall's Creek for a fuel top-up reaching Argyle for lunch.

I am in my twenty-sixth year as a driver with Gascoyne. As a child, I drove the truck on my parents' farm, starting before I could hardly walk. When the farm was sold, I worked in the south-west of the state carting logs for a couple of years. Then I joined Gascoyne and I have been there ever since. As far as accidents are concerned, 'it has all been pretty smooth sailing' apart from one rollover. On a narrow road, my co-driver hit a culvert and lost control of the vehicle. Fortunately, I only broke one bone in the rollover, a bone in my foot. Another time a co-driver lost a dog trailer on a detour. From what I have seen, any accidents are caused by inexperience, not fatigue.

In the time I have been driving, trucks have changed quite a bit; they are a lot more comfortable these days. The power today is significantly greater and, of course, we have air-conditioned cabs which we did not have when I first started with Gascoyne. In those days, I drove a Foden, capable of pulling only one trailer with a maximum speed of 70 kph 'downhill with a tail wind'. Mostly I drove between Carnarvon and Perth carting produce to Perth and general freight from Perth to Carnarvon.

Of truck drivers on the road 99 per cent are responsible. Car drivers present truck drivers with major hazards. But over the past couple of years car drivers and car drivers pulling caravans are 'becoming better educated'. The worst thing they do is see a truck coming and pull out only half of kilometre in front of you. They are parked in a parking bay and have to be in front. Often the car and caravan is travelling at 70 kph, slower than the road train, which means we in the road train have to pass them anyway.

Sleeping in a sleeper cab is something you grow used to over the years. Of the blokes I've travelled with, 99 per cent do not have any problem sleeping, especially if you have a good co-driver who thinks of you when you are sleeping and gives you the best ride he can. I can come home and sleep twelve hours straight. It is something you become used to. For 15 years, I always went to work on the same day, leaving Perth on a Tuesday and returning home on a Saturday or a Sunday. Being in a regular pattern I could come home and sleep right through the whole night. When I changed to a Friday run, for the first couple of weeks it did make a difference to my sleeping but now I am used to that routine.

As far as sleeping goes, I have no problems at all. And I do not seem to have any problem with fatigue. I rest in the truck. I can do a day's work before I leave. It is very comfortable in the truck. You can put the flap down on the bunk and it is just like being in the middle of the night when it is really the middle of the day. The sleeper cab is separately air-conditioned from the driving cab so when you are sleeping you can decide to have the air-conditioner on or off. There must be some fatigue for people when they are working. It does not matter whether you are sitting behind a desk or driving a truck. But I do not think it is a problem because there is no pressure on you as a driver. If you are a single driver, you can always pull up for a rest. If you are two-up driving, you can always swap with your co-driver.

I was a single driver for twelve years on the Port Hedland run. I had 36 hours to drive there from Perth, that is after loading you left Perth on Tuesday evening and arrived Thursday morning. It takes 24 hours driving time so you have 12 hours to sleep. You can have six hours sleep at Dongara on the first stop and stop again for another six hours the following night and still be in Hedland in time ready to unload. There is no real pressure. There is no need to take stimulants. If you need to take stimulants when you are driving two-up, then it is time to give the game away. And I know of nobody who takes stimulants who is a single driver. I do not think it is a great problem in this state.

I do not see how driving regulations could work. When it comes to driving everybody is different and has a different metabolism. Some drivers stop after four hours, others go longer – it is up to the individual. Around sunrise, for some reason, seems to be the most difficult time to drive. Back in the old days, on the Carnarvon run, the older drivers would schedule it so they would not have to drive the early morning, leaving it to the younger blokes. Perhaps it stemmed from differences between young and old drivers or, perhaps it is psychological. With long-distance driving, it is experience that counts. Many drivers drive well into their fifties. To get a driving job at Gascoyne you have to wait for someone to retire or die.

It is not only about driving–being a driver. You have to be able to get along with the company's clients. There is a lot of 'public relations' required. The clients judge the company by the driver. If they do not like the driver they will not want the truck there.

① Nor are the regulations based on empirical data –

17 Methodological issues in driver fatigue research

Ivan D. Brown

17.1 Introduction

As Lauber and Kayten (1988) have pointed out, 'the true incidence of fatigue as a causal or contributory factor (in transportation accidents) is largely unknown'. Some investigators, e.g. O'Hanlon (1978), McDonald (1984) and Storie (1984), conclude that around 10 per cent of road accidents may be attributable to fatigue, as indexed by the driver being asleep as the wheel. These and other authors suggest that fatigue is the principal contributory factor in at least 25 per cent of single-vehicle accidents. An earlier review of the problem by Harris and Mackie (1972) reported research showing that 39 per cent of commercial vehicle accidents were attributable to the driver being either asleep at the wheel or inattentive and that these categories of accident accounted for 48 per cent of the fatalities. Clearly, driver fatigue remains a problem which is extremely costly in both financial and human terms.

Why is this? The literature shows that fatigue, in general, has been researched fairly systematically for around 80 years; yet the findings from this vast body of work have still not been translated into effective countermeasures against fatigue among drivers.

One reason, of course, is identified by Lauber and Kayten (1988), who point out that 'unlike metal fatigue, human fatigue leaves no telltale signs and we can only infer its presence from circumstantial evidence'. Hence accident reports will tend to underrepresent the seriousness of the problem. In fact, as Brown (1994) notes, 'Official accident statistics may attribute as few as two per cent of road accidents to fatigue.' On this evidence, the problem appears relatively unimportant and it is hardly surprising that limited resources for safety measures are directed at more identifiable causes of accidents. A second reason may be the common but erroneous view that fatigue can be countered by the exercise of willpower, that exhortation and attitude change are the principal appropriate countermeasures and that drivers who succumb to the effects of fatigue are simply guilty of recklessness in continuing their journey while knowingly impaired. This view conveniently shifts the responsibility for avoiding fatigue-related accidents onto the individual driver, ignoring the probability that fatigue will impair not only drivers' perception and judgement of road and traffic hazards, but also their perception and judgement of their own declining abilities to cope with the driving task. A third reason, in some countries at least (O'Hanlon, 1981; Brown, 1982), is the lack of political will to introduce research-supported constraints on professional drivers' hours of work, because of the perceived adverse effect on the relative competitiveness of alternative forms of transport.

In summary, driver fatigue seems to remain a problem largely because the evidence on its contributions to road accidents is not sufficiently strong and convincing to counter popular misperceptions of the phenomenon and to overcome political resistance to imposing further constraints on the current flexibility of commercial road transport operations. This lack of convincing evidence may be partly due to the paucity of good theories of driving and of fatigue, but it seems mainly due to the difficulty of researching driving fatigue. In other words, methodological difficulties seem to underlie our failure to minimize fatigue effects on road safety, in spite of the effort that has been put into researching it since the first World War. This chapter attempts to highlight some of those difficulties and identify – hopefully – more productive approaches to the problem.

17.2 A definition of fatigue

One major difficulty over the years, no doubt, has been the loose way in which fatigue has been defined, generally failing to distinguish psychological from physiological factors. Major progress was made in this respect by Bartley and Chute's (1947) categorization of all physiological consequences of prolonged activity as 'impairment', leaving the term 'fatigue' to cover all psychological aspects of the phenomenon, including decrements in performance as well as the cognition of negative personal states. More recently, it has been recognized that performance decrements index effects of fatigue on the *effectiveness* with which a task is being performed, but not the declining *efficiency* of that performance. In other words, performance decrements are not a reliable guide to the cost to the operator of sustaining and tolerating performance at a decremented level.

Considerations of this kind persuaded Nelson (1989) to conceptualize fatigue as a *declarative* state and, for operational purposes, he defines this as 'the condition the person reaches as a result of sustained activity wherein he or she *declares* being unable to continue the activity further'. In my own writing (Brown, 1993) I have regarded this as a rather extreme view and have considered fatigue to be a graded condition which commences when individuals experience a lack of confidence in their ability to continue performing the task in hand with 100 per cent efficiency. Therefore my favoured definition of fatigue is that it is 'a subjectively experienced disinclination to continue performing the task in hand because of perceived reductions in efficiency'. This appears to be a much more useful operational definition of the phenomenon than Nelson's, because it accepts that developing fatigue is measurable and thus potentially remediable before the point of complete breakdown.

17.3 Causes of driver fatigue

We have known for many years that driver fatigue is not simply a function of time spent at the wheel, or the length of the daily work-spell. In fact, time-on-task has been found to have a relatively small effect on accident risk when daily work periods are shorter than about 11 hours (Hamelin, 1987). By contrast, chronobiological research over the past 30 years or so has shown that time-of-day can have a profound effect on human efficiency, as can inadequate sleep. Hamelin (1987) has reported that, if driving hours are 'abnormal' (i.e. occurring before about 6 a.m. or after about 6 p.m.), and especially if drivers work irregular hours, the combined effects of circadian factors and sleep disruption are clearly measurable as an increase in the relative risk of accident. His findings show that accident risk between midnight and 2 a.m. may be some 2.5 times the daily average. This is consistent with an earlier finding by Harris (1977), who reported that single-vehicle accidents were almost 2.5 times greater than expected between midnight and 8 a.m., and that accidents in which the driver was dozing at the wheel occurred almost 3.5 times more often than expected. The methodological implications of these findings for future research on driver fatigue are daunting and will be returned to later.

17.4 Factors mediating fatigue

First, we should briefly consider some of the factors which are known to have important mediating influences on fatigue and which must therefore be taken into account as independent or controlled variables in any future research.

Driver's age

Driver's age seems an obvious candidate since, as Harma (1993) has pointed out, ageing is associated with changes in the nature of sleep, decreases in the amplitude of certain circadian rhythms and possible internal desynchronization of an individual's circadian rhythm, all of which are known to influence performance efficiency. Age effects on fatigue have, in fact, been apparent for many years. For example, Foret et al. (1981) found that the critical age for the appearance of adverse effects of shiftworking in industry lay somewhere between 40 and 50. More specifically, Harris and Mackie's (1972) study of long-distance truck drivers revealed relatively greater effects of fatigue among drivers over 30 years of age, compared with younger drivers; although these differences tended to show up only during the more difficult night hours.

Personality

Personality is another obvious candidate for the mediation of fatigue, given Eysenck and Eysenck's (1964) prediction of adverse effects of unchanging task and environmental conditions on the performance of sociable extraverts. However, caution is needed in extending this prediction to other conditions, such as abnormal hours of work, given Colquhoun and Condon's (1980) finding that extraverts adjust to abnormal hours of work slightly faster than introverts. The 'morningness/eveningness' distinction reported by Folkard et al. (1979), Horne and Ostberg (1976) and others, also appears to require consideration, given that the circadian rhythms in body temperature of 'morning' types tend to peak somewhat earlier than those of 'evening' types, increasing their relative alertness. Since an earlier phase position has been associated with an increased time to adjust to abnormal working hours (Costa et al., 1989), 'morning' types should take longer to make the circadian adjustment to irregular working and will probably experience an exacerbation of fatigue symptoms while doing so. They will probably experience a further exacerbation of fatigue effects because of their greater rigidity in sleeping habits.

Physical fitness

Physical fitness is another mediating factor, the empirical evidence on which has recently been reviewed by Harma (1993). The predominant mechanism here seems to be the effect of physical exercise on improved quality of sleep, increases in maximal oxygen uptake and muscle strength, with an associated increased tolerance for working abnormal and irregular hours. Harma reports that experienced fatigue can be reduced by up to 20 per cent following certain 'physical conditioning', although the precise effects of these exercise regimes on circadian functioning remains to be explored. This fitness factor may have important mediating influences on fatigue among truck drivers, in particular, given the fact that they seem especially prone to obesity (Horne, 1992), which is known to exacerbate sleep apnoea. Since apnoeics may not be fully aware of their problem, they may not recognize that they are becoming unduly sleepy during prolonged driving spells and they may get little advance warning of 'sleep attacks'.

Driving, traffic and work experience

[Handwritten annotation: Hamiltons work shows the benefit of experience to shift work]

Driving, traffic and work experience can have mediating influences on fatigue effects. The full extent of effects of driving and traffic experience have not, to my knowledge, been systematically investigated. Slightly more is known about the mediating effects of work experience. For example, Hamelin (1987) found that certain groups of truck drivers who had experienced working abnormal hours seemed better able to compensate for fatigue, as indexed by their lower accident risk, than less experienced drivers when confronted with critical road and traffic situations. Hamelin reports that as these more experienced drivers became used to working abnormal hours, they seemed able to schedule their work more efficiently than drivers with less experience of working irregular hours. The motivating factor underlying this efficient rescheduling seems similar to the 'commitment to shiftwork' effect reported by Folkard *et al.* (1979) and by Monk and Folkard (1985) as being characterized by an appropriate adjustment of sleeping habits, eating and drinking regimes, physical exercise, and so on, all of which increased the individual's tolerance to the adverse effects of working irregular hours.

17.5 Methodological considerations

This brief overview of some of the literature on fatigue studies has been very selective, but it should suffice to identify the more important methodological factors in future research, which must be taken into consideration if we want to make faster progress in countering driver fatigue. A logical development of the foregoing overview is to consider these methodological factors under three headings:

1. Definitional implications

2. Causal implications

3. Mediational implications

[Handwritten annotation: The ideal would be to test fatigue in real world setting]

Definitional implications

The definition of fatigue adopted here has a number of very obvious implications for research methodology. First, if fatigue is a declarative state in which the individual declares being unable to continue performing the task in hand with 100 per cent efficiency, does this not mean that the task being studied has to be 'the real task, the whole task, and nothing but the task', to paraphrase a legal oath? In other words, does it not rule out the use of laboratory experimentation, simulation, and part-task studies? Ideally, yes; although we do not live in an ideal world and hence a number of scientifically interesting and theoretically ideal approaches to driver fatigue research would normally be ruled out by ethical considerations. For example, I suspect that in most developed countries we cannot ethically, for research purposes, ask drivers to perform their normal task on the road until they declare themselves as being absolutely unable to continue, in spite of the fact that the onus is on professional drivers of commercial vehicles to exercise this form of constraint every working day of their lives!

Off-road studies of driver fatigue could therefore be seen as the default option. In addition to ethical objections to 'real-life' on-road studies in general, we cannot rely, for example, on the use of retrospective studies of 'real' accident data. Clearly, these are unlikely, in isolation, to further our detailed understanding of driver fatigue, as defined here, because they provide no reliable information on the declarative state of the driver, apart from the fact that fatigue, as a contributory factor in road accidents, is likely to be under-reported by drivers attempting to avoid blame. *Socially unacceptable — negligence*

However, we also cannot rely entirely on off-road 'track' studies of fatigue, because the driver subjects in such research will know that the risk of injury to themselves or others is minimal and this knowledge is likely to affect their willingness to continue driving, in spite of perceived reductions in efficiency. We can, of course, introduce behavioural rewards and punishments and manipulate pay-off matrices in an attempt to encourage pseudo-realistic road and traffic behaviour, but there must always be some doubt as to whether the declarative state of the driver is perfectly comparable under road and track conditions. This is particularly true where certain constraints on the driver's behaviour are introduced for safety reasons. For example, the track study of driver fatigue reported by Lisper *et al.* (1986), which is frequently cited as supportive evidence on the time-course of falling asleep at the wheel, included an unrealistic constraint on maximum vehicle speed by the use of a speed limiter. The extent to which such constraints affect individuals' assessments of their ability to continue driving safely remains unknown, casting doubt on the validity of some, although probably not all, resulting fatigue findings. *Simulation provides access to these areas of fatigue*

We therefore have to accept that certain effects of fatigue, for example, on perception, cognition and mood state, can only be studied systematically via laboratory experimentation and/or simulation. However, these approaches seem more appropriate to hypothesis testing in parallel with on-road studies, than to specific research on accident risk, or the development of practical countermeasures against driver fatigue. Experience of the limited extent to which traffic authorities are prepared to translate laboratory findings into traffic legislation against fatigue suggests that such findings have low credibility, particularly when they point to the need for additional legislative constraints on individual freedom on the road, or to restrictions on the commercial competitiveness of transport operations. Nevertheless, laboratory experimentation and driving simulation must form part of the iterative approach by which research advances theory and application in this field. *As the means of advancing theory & application*

A second apparent implication of the present definition of fatigue for research methodology is that we need to concentrate only on the measurement of cognitive states, since performance measures will only index effectiveness, not perceived efficiency, and physiological symptoms will merely index impairment, which is distinct from and not necessarily correlated with declarative fatigue. I reject this implication, partly because of uncertainty over the contribution of cognition to performance change. For example; 'lane drifting' has frequently been reported as a symptom of driver drowsiness, yet there is no empirical evidence on the extent to which this is simply a result of intermittent eye-closure as compared with a conscious relaxation of steering criteria aimed at reducing task demands. We therefore cannot be absolutely sure whether lane-drifting indexes performance impairment or adaptation to a perceived reduction in information processing capability, or both. Furthermore, this implication that non-cognitive issues are unimportant, if accepted, would beg the question as to the basis of the individual's declaration of fatigue. Clearly perceptions of physiological symptoms and feedback on imperfect actions will provide individuals with two very important indications of their ability to continue performing their task competently and safely. Neither source of information may correlate with declared fatigue, since it is capable of being misperceived or ignored, particularly when the task in question is demanding. It may also not correlate with declared fatigue because it is used by individuals as an indication of the need to change their strategic and/or tactical approach to the task in hand. For example, Sperandio (1978) and Welford (1978) have identified mechanisms of this kind, by which operators implement strategy changes in order to reduce the mental load imposed by their task demands. But rather than concluding that performance and impairment measures are uncorrelated with declared fatigue and can therefore be safely ignored, we should accept that these measures can provide useful information on the nature, validity and reliability of fatigue, as declared by the individual in question. The question is, and I think it remains an empirical question, which measures of performance and of impairment provide drivers and researchers with the best guide to the development of fatigue, since we usually have to be very selective in this respect, whether conducting laboratory or field studies.

The answer to this question is not independent of the method adopted for measurement of the principal dependent variable, i.e. declarative fatigue. Approaches here range from a simple 'Yes/No' type of response to the question 'Can you continue?', to more sophisticated questionnaire studies of the kind reported by, for example, Angus and Heslegrave (1985). The actual approach chosen for studies of driver fatigue will usually represent a compromise between the need for infrequent and brief on-line interventions, in order to obtain valid data but not disrupt the driver's task, and comprehensive off-line collections of data which have no disruptive effects on driving, but which may less validly reflect fatigue as experienced at the wheel. As Morgan and Pitts (1985) have pointed out, the timing and frequency of measurement are crucial in any study of sustained operations and this will be particularly important in the study of driving, where perceived fatigue is likely to vary from moment to moment as a function of road and traffic demands. With this in mind, the validity of the data seems likely to be maximized by brief, relatively frequent and randomly scheduled on-line collections of declarative fatigue measures, plus the continuous measurement of vehicle control performance (e.g. lane-tracking, headway, acceleration, etc.) and physiological indices (e.g. heart-rate variability, GSR, EMG, etc.), with minimal off-line measures including, say, introspective reports on specific aspects of driving, such as risky overtaking.

A brief word on the subsidiary task approach to driver fatigue measurement seems relevant here since, if subsidiary task performance is a measure of reserve capacity, it could be predicted to index the fatigued driver's declining ability to cope. A variety of studies during the 1960s and 1970s (Brown, 1965; 1966) provided underwhelming support for this prediction; an outcome which I attributed mainly to the effect of additional stimulation on driver arousal (Brown, 1968; 1978). While the subsidiary task approach may not be dead as a method of measuring driver fatigue, it seems clear that any subsidiary task used for this purpose needs to be much better integrated into the driving task than has been customary in the past.

A further brief word seems relevant on the question of perceived risk, as a function of driver fatigue. If declarative fatigue reflects the driver's perceived increasing inability to cope adequately with the task, it seems plausible that this progressive uncertainty may be paralleled by an increasing perception of riskiness in continuing to drive. Now

[Risk defined] subjective risk has been defined (Brown, 1989) as a joint function of perceived hazard and self-assessed ability to cope with that hazard. These two functions therefore appear to comprise particularly interesting dependent variables for tracking fatigue onset and development, especially as their veridicality is potentially testable. In other words, their measurement would provide insight not only into the time-course of subjective risk, as the driver accumulated fatigue, but also into the specific impairment of perception, decision making and judgement. The veridicality of fatigue effects on subjective risk could be tested by investigating its correlation with objective risk, as indexed by misperceived hazards and performance impairments.

Finally, in this section, it seems relevant and timely to mention briefly the potential implications for driver fatigue methodology of upcoming technological systems for driver support. Some five years ago, the Commission of the European Communities launched its comprehensive programme of research and development called 'Dedicated Road Infrastructure for Vehicle Safety in Europe' (DRIVE), later referred to as the 'Advanced Transport Telematics' (ATT) programme (Commission of the European Communities, 1992). Based on the use of information technology linking in-vehicle and on-road systems, a number of projects within this programme aim continuously to monitor driver behaviour and assess its relevance and appropriateness to task and environmental demands. Such systems have the potential to detect the onset and progress of fatigue effects on interactions between the driver and the vehicle, the road and other road users, and either introduce warnings or guidance, or assume total control of the situation, when imminent threats to safety are recognized. If we assume that the development of reliable and affordable systems of this kind will take about 10 years and that it will be a further 10 years before the majority of vehicles are fitted with them, accidents attributable to driver fatigue could virtually be eliminated within 20 years.

The general implication of this ATT programme for driver fatigue methodology is that there is a continuing need to focus on the dependent variables of driver performance and impairment, because changes in these variables are likely to be the main source of information on driver status for the sensory input to ATT systems. Vocal and manual inputs to these systems by drivers, based on their declarative status are, of course, quite possible, but it seems unlikely that acceptable technological solutions to the problem of driver fatigue will be based solely on drivers' self-assessments. Risk homeostasis theory (Wilde, 1988) predicts that effective driver support systems may encourage drivers to offset the potential benefits these systems offer by behaving more riskily. If this prediction were fulfilled, drivers' self-assessments of their fatigued state would clearly be suspect. Hence future research will still need to identify valid and sensitive indices of fatigue within driving performance and driver impairment, and especially those indices which can readily be sensed and evaluated by upcoming technology.

Causal implications [Problems with legislation]
The methodological implications of fatigue causation have been recognized for many years. Time-on-task was the principal consideration in early studies of industrial fatigue (Vernon, 1921) and this remains the basis of current legislation on commercial drivers' hours of work. For on-road research, the main difficulty with the time-on-task variable is that it is confounded with diurnal variations in traffic demand. Attempting to deal with this problem by running 'before-and after' driving tests at the same time of day, as I have done (Brown, 1965), may serve to unconfound these two variables if one can avoid the introduction of fatigue effects transferred to driving from other sources.

Perhaps a better solution, adopted by Hamelin (1987) for example, in his study of lorry drivers' accident involvement, is to derive the dependent variable of risk rate relative to the number of drivers exposed to risk at different times of day.

The contribution of inadequate sleep to driver fatigue can obviously not be investigated validly by retrospective studies of accident involvement, except perhaps by the use of extremely costly 'on-the-spot' types of transport research. Even then, the reliability of subjects' introspections on prior nights' sleep duration must remain suspect. Any on-the-road experimental study of sleep adequacy will raise ethical questions pertaining to risk exposure. Hence the only possible approach we are left with is laboratory experimentation. This must raise the question of the purpose of further specific studies of sleep in relation to driver fatigue, given the now voluminous literature on sleep-related effects in general.

One issue which seems of theoretical and, particularly, practical interest in this respect is drivers' awareness of falling asleep at the wheel. Its practical importance stems from the legal position that drivers who realize they are about to fall asleep, yet nevertheless continue their journey, are guilty of dangerous driving. Laboratory and off-road evidence on this point appears unequivocal, but there are reasons to feel that it should remain an empirical question. For example, Lisper et al. (1986) reported an off-road study in which subjects drove a car until they fell asleep three times. An observer recorded these sleep incidents and the subjects were also required to report their occurrence. The authors' general conclusion was that drivers who fall asleep at the wheel always have to fight sleep for some considerable time before dropping off; therefore they must definitely be aware that they are at risk of falling asleep. These authors cite other German research which supports the view that it is virtually impossible to fall asleep while driving without any warning whatsoever. However, I have recently queried this conclusion (Brown, 1993), pointing to the report by Lisper et al. that some of their subjects admitted, with hindsight, having probably fallen asleep on occasions during the experiment unobserved by the experimenter and unannounced by themselves. This suggests that they may have been uncertain at the time whether they were about to fall, or had fallen, asleep.

Now we know that drowsy individuals usually experience brief and repeated sleep incidents before they finally drop off completely These 'micro-sleeps' are likely to result in detectable impairments of driving performance, such as lane-drifting, and this feedback should eventually alert drivers to their drowsy state. But are drivers always aware of the first occasion on which a micro-sleep incident occurs? At this point in fatigue onset they will have had no feedback from lane-drifting, or other symptoms of behavioural impairment, and their drowsy state will be identifiable only by the usual symptoms of tiredness, such as difficulty in focusing, increased blinking, narrowing of the visual field, muscular discomfort, and so on. Yet Lisper et al. (1986) found that such ubiquitous symptoms were reported almost equally in their study by drivers who fell asleep and those who did not; in other words, these symptoms do not provide drivers with an infallible guide to the risk of falling asleep at the wheel. I therefore suspect that fatigued drivers may occasionally be taken completely by surprise when their first micro-sleep incident occurs, although I know of no empirical evidence on this point.

Exploring the falling asleep process thus seems important, in relation to driver fatigue, and this clearly demands a psycho-physiological methodology. This approach should provide the opportunity to examine another related issue: the phenomenon referred to by the media as 'highway hypnosis' and by aviation physicians as 'empty field myopia', but which has more recently been termed 'driving without awareness' (DWA)

(Brown, 1991). The research question here is whether DWA is a separate phenomenon, distinguishable from sleep, or whether it is simply a stage in the falling asleep process? Again, this question has important practical implications relating to legal culpability in road accidents.

Circadian factors in fatigue causation have complex implications for research methodology, many of which I am not qualified to discuss in great detail, although the general implications for control of time-of-day factors in driver fatigue studies are clear. I shall mention one finding which seems relatively important here. While human performance is certainly a partial function of circadian biological rhythms, we have evidence that there is not a single 'performance rhythm' but many (Folkard and Monk, 1985). Different types of performance seem to adjust to abnormal working hours in very different ways (Craig et al., 1987). This appears important in relation to the declarative state of fatigue. If fatigued drivers are increasingly disinclined to continue their task because of perceived reductions in their efficiency of performance, which aspects of performance do they tend to use as feedback on fatigue effects? If circadian factors affect component driver skills differentially, is there not a risk that drivers may seek feedback from a component skill which actually misrepresents their current fatigued status and perhaps misleads them into thinking they are more, or alternatively less, competent to continue driving than is actually the case? I know of no evidence on this issue, although it has clear implications for the prevention of fatigue-related accidents.

Mediational implications

Age

The importance of age as a mediating factor in driver fatigue research has been enhanced by the findings mentioned earlier on its relationships with sleep and circadian rhythms. Clearly, it remains necessary to control for age in any fatigue study. However, the importance of driver age *per se* in fatigue-related road accidents is uncertain. A recent study by Summala and Mikkola (1994) did, indeed, show that recorded road accidents among 18- to 20-year-old car drivers peaked between midnight and 6 a.m., whereas among drivers over 50 years of age the peak occurred during the late afternoon hours. However, these authors suggest that the finding may be attributable to differential risk exposure, since their younger drivers continued to drive for too long at night, whereas their older drivers continued to drive for too long in the afternoon. In other words; not stopping for a rest when fatigue symptoms appear is a problem shared by all age groups. In methodological terms, the age factor thus seems more important for retrospective studies of accident rates than for experimental investigations of fatigue.

Personality

A similar comment could be made about the importance of personality for driver fatigue research, although this characteristic must obviously be taken into account in any study involving circadian factors as dependent variables because of its observed association with the adjustments of biological rhythms and sleeping habits mentioned earlier. One question which remains to be explored, to my knowledge, is the extent to which personality influences the professional driver's choice of career and persistence in the job. For example: are long-distance truck drivers predominantly introvert? If not, is there any strong association between personality and accident liability among this professional group? The answer could have important implications for driver selection and other approaches to accident reduction.

Physical fitness

The same could be said about physical fitness, although this individual characteristic is obviously a partial function of occupational hazards. Nevertheless, it remains an important consideration for methodology in driver fatigue research, especially given findings by Pleban et al. (1985) showing that fitness moderates not only cognitive work capacity, but also fatigue onset during sustained operations. Together with the problem that obesity is an occupational hazard among truck drivers and that obesity predisposes them towards obstructive sleep apnoea, thus increasing the risk of their falling asleep at the wheel, fitness is clearly an important variable, particularly in any retrospective or prospective on-road fatigue study.

Experience

Experience is a more difficult mediating factor to handle in driver fatigue research, since the evidence suggests that it is not simply vehicle handling, road and traffic experience that is important, but experience in coming to terms with particularly demanding schedules of transport operation (Hamelin, 1987). Taking this factor into account may not present difficulties in on-road fatigue studies among professional drivers, but it could prove time-consuming and costly in more experimental investigations because of the need to give subjects sufficient experience with the work-schedules being investigated.

17.6 Summary and conclusions

1. Methodological shortcomings appear to have limited the credibility of research findings on driver fatigue which, together with a lack of political will in some countries to constrain individual freedom and commercial competitiveness, has failed to minimize the contribution of driver fatigue to road accidents.

2. Defining fatigue as a declarative state, based upon self-perceptions of body distress, mood and performance, rather than defining it in terms of objective performance decrements and physiological impairment, offers more scope for progress in understanding driver fatigue and particularly in the development of practical countermeasures against the adverse effects of fatigue on road safety. This definition does cast doubt on the validity of off-road studies of driver fatigue, but these will remain the default option because of ethical restrictions on fatiguing drivers in traffic for research purposes. The definition also suggests that driver fatigue research should concentrate on purely cognitive issues. However, driving performance and physiological impairment must be of continuing interest because changes in these variables provide drivers with feedback on their declining efficiency and hence comprise an input to their declarative status. Ongoing European programmes for research and development of advanced transport telematics (ATT) will also continue to focus attention on fatigue indices other than the driver's declarative state, because overt behavioural and performance indicators of fatigue provide essential sensory inputs to many of these technology-based driver support systems.

3. The interactive causal contributions to fatigue, in order of importance, are

inadequate sleep, circadian factors and time-on-task. Their interacting effects are complex and have important implications for research methodology. Inadequate sleep and circadian factors are of primary importance in driver fatigue research, although non-driving-specific studies within the field of chronobiology should be a major source of information on the important implications of fatigue for driving.

4. Age, personality, physical fitness and specific work experience appear to be the principal factors mediating the accumulation of driver fatigue. The implications of personality for methodology in driver fatigue research seem particularly important, although fitness may also have subtle implications for fatigue as a declarative state.

5. In conclusion, much remains to be researched specifically on driver fatigue, in spite of the long history of research into fatigue in general. Three principal research questions arise from the approach taken in this chapter:

   (a) How does fatigue progressively and differentially affect drivers' self-assessments of their coping abilities and their perceptions of road and traffic hazards (with implications for misperception of subjective risk and of the cognitive and physiological costs of continuing to drive)?
   (b) Which specific decrements in task performance and which specific symptoms of physiological impairment do drivers use as feedback of their fatigued state, i.e. what is the behavioural basis of declarative fatigue?
   (c) Are drivers reliably capable of assessing the probability of their falling asleep at the wheel, before perceptible performance decrements, such as lane-drifting, have occurred (with implications for legal culpability and self-monitoring approaches to fatigue countermeasures)?

References

Angus, R.G. and Heslegrave, R.J., 1985, Effects of sleep loss on sustained cognitive performance during a command and control simulation, *Behavior Research Methods, Instruments, and Computers*, **17**, 55–67.
Bartley, S.H. and Chute, E., 1947, *Fatigue and Impairment in Man*, New York: McGraw-Hill.
Brown, I.D., 1965, A comparison of two subsidiary tasks used to measure fatigue in car drivers, *Ergonomics*, **8**, 467–74.
Brown, I.D., 1966, Effects of prolonged driving upon driving skill and performance of a subsidiary task, *Industrial Medicine and Surgery*, **35**, 760–5.
Brown, I.D., 1968, Criticisms of time-sharing techniques for the measurement of perceptual-motor difficulty, in *Proceedings of the 16th International Congress of Applied Psychology*, Amsterdam: Swets & Zeitlinger.
Brown, I.D., 1978, Dual-task methods of assessing workload, *Ergonomics*, **21**, 221–4.
Brown, I.D., 1982, Driving fatigue, *Endeavour*, **6**, 83–90.
Brown, I.D., 1989, *How Can We Train Safe Driving?* Groningen, The Netherlands: University of Groningen Traffic Research Centre.
Brown, I.D., 1991, Highway hypnosis: implications for road traffic researchers and practitioners, in Gale, A.G., Brown, I.D., Haslegrave, C.M., Moorhead, I. and Taylor, S.P. (Eds), 1991, *Vision In Vehicles III*, Amsterdam: North-Holland.
Brown, I.D., 1993, Driver fatigue and road safety, *Alcohol, Drugs and Driving*, **9**, 239–52.
Brown, I.D., 1994, Driver fatigue, *Human Factors*, **36**, 219–31.
Colquhoun, W.P. and Condon, R., 1980, Introversion-extraversion and the adjustment of body temperature rhythm to night work, in Reinberg, A., Vieux, N. and Andlaur, P. (Eds), *Night and Shiftwork: Biological and Social Aspects*, Oxford: Pergamon Press.
Commission of the European Communities, 1992, *DRIVE '92: Research and Technology Development in Advanced Transport Telematics in 1992*, Brussels: CEC.

Costa, G., Lievore, F., Casaletti, G., Gaffuri, E. and Folkard, S., 1989, Circadian characteristics influencing interindividual differences in tolerance and adjustment to shiftwork, *Ergonomics*, **32**, 373–85.

Craig, A., Davies, D.R. and Matthews, G., 1987, Diurnal variation, task characteristics and vigilance performance, *Human Factors*, **29**, 675–84.

Eysenck, H.J. and Eysenck, S.B.G., 1964, *Manual of the Eysenck Personality Inventory*, London: University of London Press.

Folkard, S. and Monk, T.H. (Eds), 1985, *Hours of Work: Temporal Factors in Work-Scheduling*, Chichester: John Wiley & Sons.

Folkard, S., Monk, T.H. and Lobban, M.C., 1979, Towards a predictive test of adjustment to shiftwork, *Ergonomics*, **22**, 79–91.

Foret, J., Bensimon, G., Benoit, O. and Vieux, N., 1981, Quality of sleep as a function of age and shiftwork, in Reinberg, A., Vieux, N. and Andlauer, P. (Eds), *Aspects of Human Efficiency*, London: English Universities Press, pp. 273–82.

Hamelin, P., 1987, Lorry drivers' time habits in work and their involvement in traffic accidents, *Ergonomics*, **30**, 1323–33.

Harma, M., 1993, Individual differences in tolerance to shiftwork: a review, *Ergonomics*, **36**, 101–9.

Harris, W., 1977, Fatigue, circadian rhythm and truck accidents, in Mackie, R.R. (Ed.), *Vigilance: Theory, Operational Performance and Physiological Correlates*, New York: Plenum Press, pp. 133–46.

Harris, W. and Mackie, R.R., 1972, A study of the relationships among fatigue, hours of service and safety of operations of truck and bus drivers, *Human Factors Research Inc.*, Report No.1727-2.

Horne, J.A., 1992, Stay awake, stay alive, *New Scientist*, **133**, No. 1802.

Horne, J.A. and Ostberg, D., 1976, A self-assessment questionnaire to determine morningness-eveningness in human circadian rhythms, *International Journal of Chronobiology*, **4**, 97–110.

Lauber, J.K. and Kayten, P.J., 1988, Sleepiness, circadian dysrhythmia and fatigue in transportation system accidents, *Sleep*, **11**, 503–12.

Lisper, H.-O., Laurell, H. and van Loon, J., 1986, Relation between time to falling asleep behind the wheel on a closed track and changes in subsidiary reaction time during prolonged driving on a motorway, *Ergonomics*, **29**, 445–53.

McDonald, N., 1984, *Fatigue, Safety and the Truck Driver*, London: Taylor & Francis.

Monk, T.H. and Folkard, S., 1985, Individual differences in shiftwork adjustment, in Folkard, S. and Monk, T.H. (Eds), *Hours of Work: Temporal Factors in Work-Scheduling*, Chichester: John Wiley & Sons, pp. 227–37.

Morgan, B.B. and Pitts, E.W., 1985, Methodological issues in the assessment of sustained performance, *Behavior Research Methods, Instruments, and Computers*, **17**(1), 96–101.

Nelson, T.M., 1989, Subjective factors related to fatigue, *Alcohol, Drugs and Driving*, **5**, 193–214.

O'Hanlon, J.F., 1978, What is the extent of the driving fatigue problem?, in *Driving Fatigue in Road Traffic Accidents*, Brussels: Commission of the European Communities Report No. EUR6065EN, 19–25.

O'Hanlon, J.F., 1981, The failure of EEC Regulation No. 543/69 to prevent serious consequences of fatigue in European commercial vehicle operations; an American's opinion, *KNVTO Nederlands Transport*, **9**, 305–9.

Pleban, R.J., Thomas, D.A. and Thompson, H.L., 1985, Physical fitness as a moderator of cognitive work capacity and fatigue onset under sustained combat-like conditions, *Behavior Research Methods, Instruments, and Computers*, **17**(1), 86–9.

Sperandio, J.-C., 1978, The regulation of working methods as a function of work-load among air traffic controllers, *Ergonomics*, **21**, 195–202.

Storie, V.J., 1984, *Involvement of goods vehicles and public service vehicles in motorway accidents*, UK Dept. of Transport/Transport and Road Research Laboratory, Report No. 1113.

Summala, H. and Mikkola, T., 1994, Fatal accidents among car and truck drivers: effects of fatigue, age, and alcohol consumption, *Human Factors*, **36**, 315–26.

Vernon, M.H., 1921, *Industrial Fatigue and Efficiency*, New York: Dutton.

Welford, A.T., 1978, Mental workload as a function of demand, capacity, strategy and skill, *Ergonomics*, **21**, 151–67.

Wilde, G.J.S., 1988, Risk homeostasis theory and traffic accidents: propositions, deductions and discussion of dissention in recent reactions, *Ergonomics*, **31**, 441–68.

18 Alcohol, fatigue and human performance deficits: insights from additive factor logic

Colin Ryan, Janet Greeley and Katherine Russo

18.1 Introduction

There is no doubt that alcohol, given in sufficient doses, affects human performance in a wide variety of perceptual-cognitive and motor tasks (Howat *et al.*, 1991). We know, also, that those effects can be exacerbated by a host of factors including task complexity, concurrent attentional demands, fatigue, etc. Regrettably, research in this area has tended to be task-predicated rather than process- or operation-predicated. In consequence, we have an extensive catalogue of tasks known to be more-or-less affected by alcohol – most of the data scarcely surprising – but little useful generalization or theoretical integration.

One alternative to this atomistic cataloguing of tasks affected by alcohol, is to focus on their constituent processes or operations – the task substrate. These underlying operations are, arguably, task-general, the building blocks of many complex skills. By establishing which component operations are susceptible to factors of interest (e.g. alcohol, fatigue, amphetamines, etc.) we advance the cause of a more general task-independent theory of human performance (Huntley, 1972; Moscowitz and Burns, 1973).

The feasibility of this approach depends, of course, on being able to parse tasks into their constituent operations, meaningfully and reliably – an enterprise which beguiled Donders as long ago as 1868. He advanced a logic and a method for decomposing reaction times into the durations of component processes which was initially adopted with some enthusiasm (Jastrow, 1890), but a succession of critics including Wundt, Ach, Kulpe and Watt (Woodworth, 1938) and later Crossman (1953) and Johnson (1955) substantially discredited Donders' approach.

The use of reaction-time (RT) measures to make inferences about implicit cognitive operations underwent a resurgence in the 1970s due principally to Saul Sternberg's (1969) exegesis of additive-factor logic and method. The additive-factor method is a technique for inferring the number and organization of cognitive operations from patterns of additivity and interaction between experimental factors. It assumes that the duration of successive stages are the additive components of reaction time: mean RT is the sum of the means of the component durations. Experimental factors (such as fatigue or alcohol) are assumed to affect the durations of some stages and not others. Factors whose effects on mean RT are additive are assumed to affect different stages, while factors which interact are held to affect at least one stage in common. Thus, a variance table can provide powerful insights into the locus of factor effects, enabling one to resolve both the number and nature of the underlying stages or operations. Sternberg developed and applied his method to data from his memory search paradigms.

Sternberg (1966) had subjects decide whether a probe item was a member of a small predefined set. This memory set was either changed on each trial (Sternberg's varied-set task) or held constant and repeatedly probed (his fixed-set task). In both cases, mean RT increased linearly with setsize at the same rate for both positive and negative responses. Sternberg interpreted these data as evidence for an underlying serial exhaustive scanning process at a rate of approximately 30 items per millisecond. A third variant of this paradigm, first used by Forrin and Morin (1969) is the so-called concurrent-sets paradigm in which fixed and varied sets are concurrently relevant and simultaneously probed.

In the study reported here, we utilized all three paradigms in order to examine the pattern of additivity and interaction between factors – in order to establish the magnitude and locus of alcohol and, ultimately, fatigue effects.

18.2 Method

Subjects

Volunteer subjects (12 female and 12 male) aged 18–41 years ($\bar{X} = 22.3$; sd $= 5.8$) were tested. All were screened by telephone prior to testing. Subjects were excluded from the study if they:

- were currently taking prescribed medication or had used non-prescription drugs during the previous month;
- were currently undergoing medical treatment; or
- had a history of psychological or psychiatric illness, or had at any time sought treatment for alcohol problems.

Additionally, smokers and heavy caffeine drinkers (more than five cups per day) were excluded from the study to obviate the possibility of confounding acute withdrawal across the extended testing session. All subjects were required to have drunk a minimum of five standard drinks in one hour on at least one previous occasion without becoming ill. They were also asked to refrain from

- eating for two hours prior to testing;
- drinking alcohol or taking other recreational drugs for 24 hours prior to testing;
- from drinking coffee for at least one hour prior to arriving at the laboratory.

Twelve subjects were randomly assigned to each of the alcohol (A) and placebo (P) groups. They were unaware they had been so assigned. All expected to receive alcohol. Post-test, group A subjects were requested to remain in the laboratory until their breath–alcohol concentrations (BACs) returned to zero. All subjects were fully informed of the nature of the task and required to sign a consent form prior to testing.

Design

The experiment consisted of three distinct testing cycles. Cycle 1 was a baseline or pretest phase; neither group had alcohol. Group A subjects ingested alcohol at the end of Cycle 1, while group P subjects received an equivalent amount of orange juice with 5 ml of vodka floated on the surface. The blood–alcohol concentrations of group A typically ascended in Cycle 2 and (following a standard light lunch) descended in Cycle 3. For clarity, we sometimes refer to these simply as the ascending BAC and descending BAC cycles, respectively.

Subjects completed 180 trials in each of the varied-set (VS), fixed-set (FS) and concurrent-sets (CS) memory scanning tasks in each cycle, a total of (180 trials × 3 tasks × 3 cycles =) 1620 trials in all. Task order was counterbalanced across subjects within groups and between groups. The probability of a target being present in the memorized positive set was held constant at 0.5 in all three tasks.

Varied set task

Positive sets consisted of 2, 4 or 6 items selected at random without replacement from the pool of digits 0–9. On positive (yes) trials, the probe item was a member of this memorized set, being drawn equally often from each serial position. On negative (no) trials, probes were selected at random from the residual (non-set) digits. Presentation was blocked by setsize and setsize-sequence was counterbalanced across subjects within groups and identical for the two groups to avoid confounding setsize and alcohol effects with practice effects. The sequence of positive and negative trials was randomized with the constraint that, within a block, each response type was equiprobable ($p = 0.5$). Thirty complete repetitions of the within-subjects factor combinations were used in each cycle ($3 \times 2 \times 30 = 180$ trials). In summary, then, a 2 × [3 × 2 × 30] design examined the effects of groups, setsize, response type and repetitions respectively on reaction time and error rates.

Fixed set task

Twenty-four letters of the alphabet were randomly assigned to positive and negative pools of 12 items for each subject. The letters O and I were excluded because of their confusability with digits 0 and 1. Nested positive sets of 2, 4, 6, 8, 10 or 12 letters were constructed from the positive pool, such that the set of 2 was included in the set of 4, the set of 4 was included in the set of 6, and so on. Within blocks, each serial position of the positive set was probed equally often. Negative probes were chosen at random, with replacement, from the negative pool. Thus, a 2 × [6 × 2 × 15] between-within design examined the effects of group, setsize, response type and repetitions on response times. The order in which setsize conditions were encountered was counterbalanced across subjects within groups, and was identical for the two groups.

Concurrent sets task

On each trial, subjects were required to decide if a probe item was a member of a newly defined varied set, or a previously learned fixed set. They responded yes if the probe was a member of either set and no if it was not. There were 2, 4, or 6 items in the varied set and 2, 4, 6, 8, 10 or 12 items in the fixed sets. The fixed sets were (again) nested. In essence, we merged the FS and VS tasks outlined above into a single task. In this case, a 2 × [3 × 6 × 2 × 5] between-within design examined the effects of group, varied setsize, fixed setsize, response type and repetitions respectively. Presentation was blocked by set sizes and the order of both fixed and varied set sizes was counterbalanced across subjects within groups and was identical for the two groups.

18.3 Apparatus

Stimulus sequences were generated and displayed, responses were recorded and stored, and RTs were measured using custom software routines driving 50 MHz 486 PCs. Responses were registered by subjects depressing the appropriate one of two adjacent keys on the keyboard numberpad. Reaction times were calculated using the computer's on-board clock. The subject's BAC was measured using a Drager Alcotest Type 7410 Breathalyser.

18.4 Procedure

Preliminaries

Prior to testing, subjects were weighed and their BACs were checked. All registered 0.00 per cent. Participants then read a general information sheet, which detailed the procedures to be followed and outlined their rights. All subsequently signed a consent form. Task-specific instructions were then given. All three paradigms were explained and demonstrated. Care was taken to ensure they fully understood each task. Subjects were asked to respond as quickly as possible consistent with virtually error-free performance. Subjects responded using the index and middle fingers of their preferred hand to depress the appropriate one of two adjacent keys on the computer keyboard numberpad. All subjects were led to expect alcohol, informed of the maximum dose which could be administered, but were told that the dose may vary across subjects.

General testing procedure

The three distinct cycles of the experiment each lasted approximately 90 minutes. There was a 20-minute break between Cycles 1 and 2, and a 45-minute break between Cycles 2 and 3. Subjects were comfortably seated in front of a computer monitor with the numberpad of the keyboard conveniently accessible to their preferred hand.

Alcohol dose and measurement

Subjects in Group A received 1.0 g/kg body weight of alcohol in the form of 37.2% alc/vol vodka diluted with fresh orange juice in the ratio of 3:1 by volume. The alcohol dose was calculated to yield, and in fact yielded, a mean peak BAC of 0.1 per cent. Group P subjects received an equivalent amount of juice only, with a small amount (5 ml) of alcohol floated on top immediately prior to serving. The BACs of both groups were taken at the end of each setsize block (in the VS and FS tasks) or at the end of each FS block (in the CS task) throughout the ascending and descending BAC cycles. Subjects were given no feedback as to their BAC. They were, however, asked to estimate their BAC each time it was measured.

Varied set task

On each trial, memory set items were presented serially, at a central, fixed screen locus, at the rate of one item per second. An asterisk displayed for one second after the last item in the positive set marked the end of the set and prepared subjects for probe onset one second after asterisk-offset. Subjects heard a tone if their response was correct and their RT in milliseconds was displayed at the fixed locus for one second, completing the trial. If the response was incorrect, no tone was presented and the word error was displayed in lieu of RT. The inter-trial interval was 1.5 seconds. Sixty trials were given at each of the three set sizes (= 180). At the end of each block, a screen message instructed subjects to rest for one minute before initiating the next block of trials.

Fixed set task

The appropriate memory set was listed on the computer monitor at the commencement of each block. Subjects were instructed to learn that list of letters to a criterion of three complete repetitions without error or hesitation. They were tested by the experimenter (E), before initiating the sequence of trials, to ensure that they had reached criterion. If they had not, the positive set was relearned. Thirty trials were given at each of the six set sizes (= 180). Each trial consisted of a one second warning asterisk followed one second after asterisk-offset by a single probe letter. Asterisk and probe were displayed at a single, central, fixed locus. Performance feedback was identical to that given in the VS task. Again, a one minute rest period was given between blocks. The ITI was 1.5 seconds.

Concurrent sets task

The fixed positive set of letters for a given block of trials was displayed at the beginning of a block, learned to the same criterion, and tested as in the simple FS task. Additionally, on each trial, subjects were serially presented with a positive varied set of digits. Display, probing and performance feedback were identical to the simple VS task. Subjects depressed the 'yes' key if the probe was in either the fixed set or the current varied set and the 'no' key if it was not a member of either. Again, the ITI was 1.5 seconds. Ten trials were given for each of the 18 setsize-combinations (6 × 3). Presentation was blocked by fixed setsize then by varied setsize.

Errors and missing data

Across tasks and cycles, subjects responded incorrectly on 11.9 per cent of trials. Errors were excluded from the principal analyses and error RTs were replaced with the mean of correct responses in that anova cell. Errors were analysed separately, but are not discussed in detail in this chapter. Repetitions *per se* were of little theoretical interest and were not analysed. All trend lines were fitted by least-squares estimation.

18.5   Results

In this section, we begin by tracking performance for each task separately across the three testing cycles.

Varied set task

Cycle 1 – Pretest

A three way split-plot anova contrasted the effects of group (placebo/alcohol), setsize ($s = 2, 4, 6$) and response type (yes/no) on mean response latencies. As might be

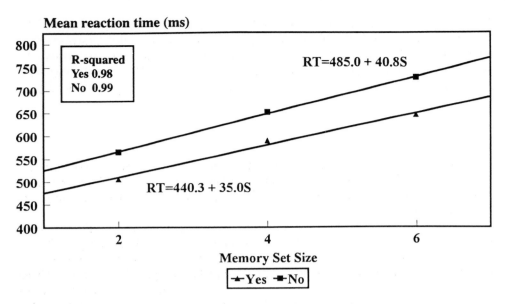

Figure 18.1 Varied set task, Cycle 1–(pre-test): mean RT as a function of setsize with response type as parameter.

expected, performance of the randomly assigned alcohol and placebo groups was equivalent in this pretest phase ($F < 1$). The effect of setsize was highly significant ($F(2, 44) = 34.72; p < 0.001$), as was response type ($F(1, 22) = 43.79; p < 0.001$). Mean RT is plotted as a function of setsize in Figure 18.1 with response type as parameter. In the absence of a significant group effect the data for all subjects are pooled.

The data are very well described by the linear functions shown. That goodness-of-

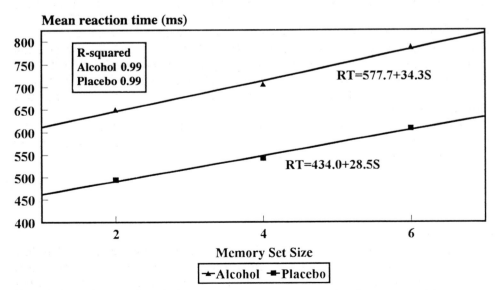

Figure 18.2 Varied set task, Cycle 2–(ascending BAC): mean RT as a function of setsize with alcohol as parameter.

fit is attested to by the very high R-squared values included in that figure. The setsize × response type interaction did not approach significance ($F < 1$), attesting to the essentially parallel 'yes' and 'no' functions. The data are, thus, quite consistent with Sternberg's serial-exhaustive scanning model of STM search. The slope estimates of 35.0 ms and 40.8 ms, respectively, accord well with those reported by Sternberg (1966). No other interactions approached significance ($Fs < 1$).

Cycle 2 – Ascending BAC

An anova analogous to that used to check the Cycle 1 data confirmed the significance of setsize ($F(2, 44) = 12.63; p < 0.001$) and response type ($F(1, 22) = 17.71; p < 0.001$). Again, there was no hint of an interaction ($F < 1$), or of a higher-order interaction with the group factor.

Serial-exhaustive scanning is the apparent modus operandi of all subjects. More important, in this context, the group effect was highly significant ($F(1, 22) = 17.43; p < 0.001$); alcohol markedly affected performance in this VS paradigm. Mean RT is plotted as a function of setsize with group as parameter in Figure 18.2. It is clear from that figure that alcohol affects the zero-intercept but not the slope of the setsize function – performance is slowed (the group A intercept is some 123.7 ms greater than the group P) but scanning rates were unaffected by the quite substantial group A alcohol dose. The group × setsize interaction did not approach significance ($F < 1$). These data are generally consistent with those reported by Maylor and Rabbitt (1989) for a similar paradigm: alcohol appears to impact on the non-scanning components of the task reflected in the zero intercept (stimulus encoding, response selection or response execution). No other interactions approached significance ($F's < 1$).

Cycle 3 – descending BAC

There was some (rather equivocal) evidence for a strategy shift in this final cycle of the VS task. Again, setsize ($F(2, 44) = 40.25; p < 0.001$) and response type ($F(1, 22) = 58.96; p < 0.001$) were highly significant. However, in this cycle, there was some indication of an interaction between these factors ($F(2, 44) = 3.051; p = 0.056$). This interaction derived from negative latency functions being steeper than positive by about 10 ms – indicative, perhaps, of a partial shift to serial, self-terminating scanning as fatigue encroached. More important, here, the effect of alcohol on VS performance was now marginal ($p = 0.10$).

The lack of any group × setsize interaction ($F < 1$) and of any significant higher order group interactions ($F's < 1$), is generally consistent with the view that the locus of alcohol effects is in the non-scanning components of this VS task.

Fixed set task

Cycle 1 – Pretest

Again, predictably, the effect of group did not approach significance in this baseline phase ($F < 1$). While there were significant effects of both setsize ($F(5, 110) = 7.61; p < 0.001$) and response type ($F(1, 22) = 8.02; p < 0.01$), their interaction did not approach significance ($F < 1$). Mean RT is plotted as a function of setsize in Figure 18.3 with response type as parameter. The positive and negative latencies are well

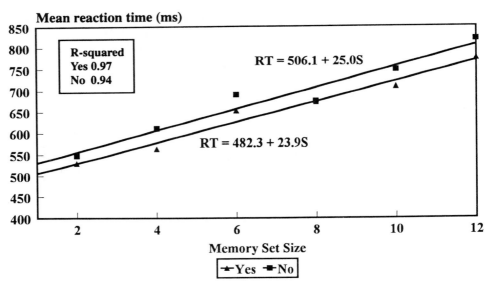

Figure 18.3 Fixed set task, Cycle 1–(pre-test): mean RT plotted as a function of setsize with response type as parameter.

described by the linear functions shown, with R^2 equal to approximately 0.97 and 0.94 respectively. As well as being convincingly linear, the functions are statistically parallel.

These data are somewhat surprising, being evidence for a serial, exhaustive, scan in supra-span lists of up to 12 items. If controlled serial scanning goes on in STM (Sternberg, 1966; Schneider and Shiffrin, 1977; Shiffrin and Schneider, 1977) then STM is, indeed, a large place about which we know relatively little (Ryan, 1983).

Cycle 2 – ascending BAC

There is a clear-cut effect of alcohol on performance in this FS paradigm ($F(1, 22) = 10.09; p < 0.01$).

Item-recognition functions for both the alcohol and placebo groups were again convincingly linear. Mean latencies for both alcohol and placebo groups are well-described by the linear trends shown. Again, setsize ($F(5, 110) = 13.02; p < 0.001$) and response type ($F(1, 22) = 12.52; p < 0.01$) are significant. Of some interest, here, is the fact that group P scanned at twice the rate of subjects who had ingested alcohol. Group A subjects were scanning at rates equivalent to those seen in the VS task. Group P subjects appeared uniquely to benefit from set fixity. These data need to be interpreted with caution in the absence of a significant groups × setsize interaction. We did, however, observe a similar pattern of data in the Cycle 3 analysis reported below: group P scanned faster than group A. Group A subjects also appeared to scan faster as BAC dropped in Cycle 3. The group × setsize interaction across the two cycles was statistically marginal ($F(5, 110) = 1.9; p = 0.10$).

Cycle 3 – descending BAC

Alcohol continued to affect performance as mean BAC diminished; group A response latencies were still significantly slower overall ($F(1, 22) = 8.19; p < 0.01$). The effects

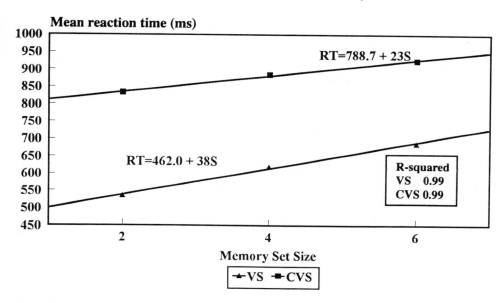

Figure 18.4 Concurrent sets task, Cycle 1–(pre-test): mean RT as a function of the number of items in the varied set in the single-task condition (VS) and the concurrent-sets paradigm (CVS).

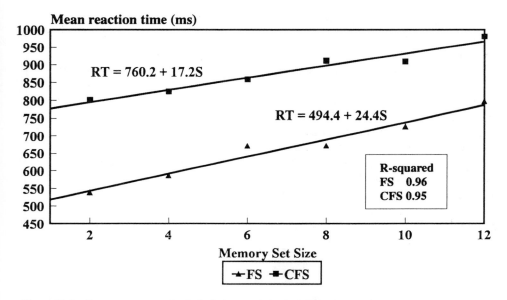

Figure 18.5 Concurrent sets task, Cycle 1–(pre-test): mean RT as a function of the number of items in the fixed set in the single-task condition (FS) and the concurrent-sets paradigm (CFS).

of setsize ($F(5, 110) = 17.60; p < 0.01$) and response type ($F(1, 22) = 32.65; p < 0.0001$) remained highly significant.

Concurrent sets task

Cycle 1 – Pretest

A four-way split-plot anova checked the effects of group, fixed setsize, varied setsize, and response type on performance. Predictably, there was no difference between groups in this pretest phase ($F < 1$). While both fixed setsize ($F(5, 110) = 5.31; p < 0.001$) and varied setsize ($F(2, 44) = 6.41; p < 0.01$) significantly affected response latencies, they did not interact – the tasks appear to run off independently. Both FS and VS functions were quite well-described by the linear functions shown, and neither fixed setsize nor varied setsize interacted with response type (F's < 1), data generally consistent with independent serial exhaustive scans of both sets. There was, however, a marked cost of concurrency *per se*. Those costs are summarized in Figures 18.4 and 18.5, in which the baseline (single-task) VS and FS data are contrasted with equivalent functions in the CS condition.

Cycle 2 – ascending BAC

Mean RT is plotted as a function of varied setsize in Figure 18.6, and fixed setsize in Figure 18.7, with groups as parameter. The effect of groups was significant ($F(1, 22) = 7.36; p < 0.05$) – performance is again markedly slowed in group A. Both varied setsize ($F(2, 44) = 9.99; p < 0.001$) and fixed setsize ($F(5, 110) = 5.82; p < 0.001$)

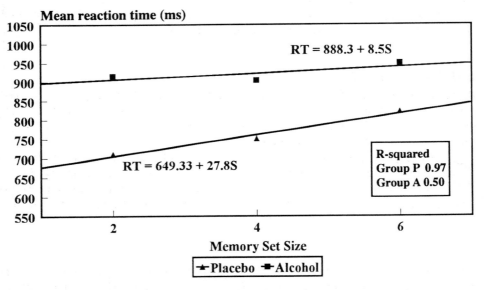

Figure 18.6 Concurrent sets task, Cycle 2–(ascending BAC): mean RT as a function of varied setsize with alcohol as parameter.

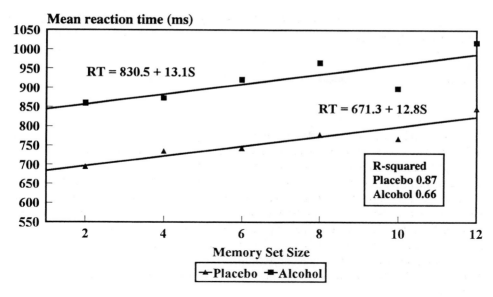

Figure 18.7 Concurrent sets task (CFS), Cycle 2–(ascending BAC): mean RT as a function of fixed setsize with alcohol as parameter.

significantly affected mean RT, and there was no hint of their interaction or any significant higher order interaction ($F's < 1$).

Most noticeable is the exceedingly shallow slope (8.5 ms/item) of the group A function in Figure 18.6 (suggesting a shift to probe categorization from serial scanning – on a substantial proportion of trials at least), while group P subjects continued with a standard serial exhaustive scan. The searches remained effectively independent, even when subjects had received a substantial dose of alcohol.

There is a suggestion in Figure 18.7 that both group A and group P subjects are probabilistically mixing serial scanning and probe categorization.

Cycle 3 – descending BAC

Alcohol continued to affect performance significantly as BACs fell ($F(1, 22) = 8.44$; $p < 0.01$) and a number of previously unseen effects began to emerge, almost certainly due to interactions between the complexity of the task, residual effects of alcohol, and the advanced fatigue of many subjects.

Varied setsize was no longer significant ($F < 1$) and there was some evidence for the independence of FS and VS searches beginning to break down: the observed interaction between fixed and varied set sizes and between those set sizes and groups increased, while remaining strictly non-significant ($p = 0.09$ and 0.07 respectively).

Errors

Errors per cent is plotted on the abscissa of all figures. Alcohol significantly increased errors rates in the ascending BAC phase (Cycle 2) of all three tasks ($p < 0.05$); misses increased sharply, while false alarm rates were unaffected by alcohol. The increase in misses independent of changes in the false alarm rates is consistent with an inability to retain positive set items with alcohol on board, rather than a criterion shift *per se*. Between-groups differences in error rate were not significant in Cycle 3 of any of the three tasks ($p > 0.05$).

18.6 Discussion

The present single-task VS data confirm Maylor and Rabbitt's (1989) suggestion that alcohol affects operations other than the serial scanning process, namely those encoded in the zero intercept of item-recognition functions (stimulus preprocessing, response selection, response execution). There is some evidence, from both the FS and CS tasks, that alcohol may also occasion shifts in scanning strategy. Independent and parallel searches of the fixed and varied sets in evidence in the CS pretest (Cycle 1) tend to break down as a consequence of fatigue and alcohol, alone and in combination. Several aspects of the data have interesting implications for Schneider and Shiffrin's (1977; Shiffrin and Schneider, 1977) theory of automaticity and control. In Cycle 1 of the CS task, there is clear evidence for concurrent, non-interfering serial, controlled, searches of the fixed and varied sets. According to Shiffrin and Schneider this is impossible; concurrent controlled processes must interfere – by definition (Ryan, 1983). In the same vein, the baseline Cycle 1 FS and CS data indicate controlled serial scanning in decidedly supra-span lists of up to twelve items. Since controlled processes go on in STM – by definition – these data are highly problematic for any unelaborated version of Shiffrin and Schneider's automaticity-control distinction (Ryan, 1983).

In a recent meta-analysis of RT and error data, Maylor and Rabbitt (1993) have argued that alcohol *globally* slows performance and, accordingly, that alcohol-induced deficits are not usefully described as stage-specific: RT with alcohol on board is a constant multiplicative proportion of RT without alcohol. Their global slowing hypothesis entails, then, that any experimental factor which affects reaction time must interact with alcohol. The present data provide strong evidence against that notion: in the VS and FS tasks, for example, both setsize and response type have marked effects on RT throughout. Neither factor shows any hint of a significant interaction with alcohol ($F < 1$ in all cases).

References

Crossman, E.F.R.W., 1953, Entropy and choice time: the effect of frequency unbalance on choice responses, *Quarterly Journal of Experimental Psychology*, 5, 41–51.
Forrin, B. and Morin, R.E., 1969, Recognition times for items in short- and long-term memory, *Acta Psychologica*, 30, 126–41.
Howat, P., Sleet, D. and Smith, I., 1991, Alcohol and driving: is the 0.05% blood alcohol concentration limit justified? *Drug and Alcohol Review*, 10, 151–66.
Huntley, M.S., 1972, Influences of alcohol and S-R uncertainty upon spatial localisation time, *Psychopharmacologia*, 27, 131–40.
Jastrow, J., 1890, *The Time-Relations of Mental Phenomena*, New York: N.D.C. Hodges.
Johnson, D.M., 1955, *The Psychology of Thought and Judgement*, New York: Harper.
Maylor, E.A. and Rabbitt, P.M.A., 1989, Relationship between rate of preparation for, and processing of, an event requiring a choice response, *The Quarterly Journal of Experimental Psychology*, 41, 47–62.
Moscowitz, H. and Burns, M., 1973, Alcohol effects on information processing time with an overlearned task, *Perception & Motor Skills*, 37, 835–9.

Ryan, C.C., 1983, Reassessing the automaticity-control distinction: item recognition as a paradigm case, *Psychological Review*, **90**, 171–8.

Schneider, W. and Shiffrin, R.M., 1977, Controlled and automatic human information processing: I. Detection, search and attention, *Psychological Review*, **84**, 1–66.

Shiffrin, R.M. and Schneider, W., 1977, Controlled and automatic human information processing: II. Perceptual learning, automatic attending, and a general theory, *Psychological Review*, **84**, 127–90.

Sternberg, S., 1966, High speed scanning in human memory, *Science*, **153**, 652–4.

Sternberg, S., 1969, Memory scanning: mental processes revealed by reaction-time experiments, *American Scientist*, **57**, 421–57.

Woodworth, R.S., 1938, *Experimental Psychology*, New York: Holt.

19 Driver impairment monitoring by physiological measures

Karel Brookhuis

19.1 Introduction

It has been argued that prevention or reduction of traffic accidents requires understanding the mechanisms of accident causation (Brown, 1991). The identification of specific (negative) behaviours contributing to accidents is vital for this understanding. Specific negative driving behaviours, as a part of driving performance in general, are subject to the same factors that affect human performance in the sense that driver impairment could be related to these behaviours. Therefore, in order to identify in particular driver impairment preceding the insidious behaviours that lead to increasing accident risk, combined continuous monitoring of driving behaviour and driver's physiology seems potentially contributing to traffic safety.

It has also been argued that driving performance easily deteriorates, for instance as a consequence of use of alcohol or medicinal drugs, illness, sleep deprivation, temperature, driver underload or overload (Brookhuis et al., 1991); for an overview see Thomas et al. (1989). These factors are all known to affect driver state in the sense that they are causatively related to driver impairment, and thus causatively related to performance decrement. Changes in driver state are reflected in changes in relevant physiological parameters such as electro-encephalogram (EEG), electrocardiogram (ECG), galvanic skin response (GSR), electromyogram (EMG), etc. (Brookhuis and de Waard, 1993). Ideally, in order to estimate the contribution of the changes in driver state to accident causation, one would want to be able to measure these physiological signals continuously in real-life driving. However, it is highly unlikely that measuring physiological parameters while driving a motor vehicle will ever become standard. Therefore, measuring physiological parameters could also be considered from the point of view of validating non-obtrusive in-vehicle measures that might be used to monitor driver state continuously through vehicle parameters.

There is a considerable literature about the relationship between vehicle parameters and generally occurring changes in driver state from different sources (Thomas et al., 1989). Driver state is not some unitary, fixed phenomenon, not even within an individual. It varies with time of day, age, subjective feelings (mood), but also with time-on-task and all kinds of external influences, such as traffic environment and situational task-load, alcohol and (medicinal) drugs.

One approach of research in this field could be to locate the optimum with respect to driving performance, and then to detect signs of deviations from the optimum (Wiener et al., 1984), first from the individual's physiology, being the forebode of behaviour, then from the relevant behavioural parameters themselves. To assess an individual's optimum and deviations from that optimum with respect to the relationship between physiology and performance, careful research should be carried out under controlled circumstances. Having established this relationship, it should be possible to predict deviations from optimal behaviour in the actual driving environment.

19.2 Alcohol and drugs

Alcohol and drug consumption may serve as first examples of sources of changes in driver state. It is well known, i.e. widely accepted, that alcohol and, probably to a lesser degree, drugs have a detrimental effect on driving performance (Seppala *et al.*, 1979; Smiley and Brookhuis, 1987). Epidemiological studies on alcohol and accident involvement have made a strong case for the alcohol part of this assertion. A relationship between the amount of alcohol and the amount of weaving has been established in the actual driving environment by Louwerens *et al.* (1987), who reported that performance started to deviate from 'baseline' at 0.06 per cent BAC. Laboratory studies showed impairments of driving-related skills after alcohol or drug intake. Especially reaction time and detection performance deteriorate as a function of alcohol dose or degree of sedation by (mostly medicinal) drugs. Relatively low levels of alcohol impaired sensory-perceptual tasks more than psychomotor tasks (Levine *et al.*, 1973). Weaving belongs largely to the latter category, which leads to the assertion that a field test of sensitivity for alcohol effects must not be based exclusively on this weaving test, but also developed along the lines of reaction time tests including sensory-perceptual variables (Smiley and Brookhuis, 1987; Brookhuis *et al.*, 1987). The impression might otherwise arise that the effects of alcohol on driving performance (and traffic safety in general) are not so negative below 0.06 per cent BAC. In accordance with this assertion, De Waard and Brookhuis (1991) found that at a BAC just below the 0.05 per cent legal limit in The Netherlands, i.e. between 0.046 per cent and 0.035 per cent during the experiment, subjects responded more slowly to speed changes of a leading car; alcohol resulted in an impairment of 19 per cent compared to baseline delay. However, in the same experiment De Waard and Brookhuis found an increase in the standard deviation of the lateral position of more than 5 cm, i.e. 30 per cent impairment, whereas Louwerens *et al.* (1987) reported such an impairment after a BAC over 0.10 per cent.

A number of physiological changes have been reported as a result of alcohol and drug intake. Brookhuis *et al.* (1986) found increased energy in alpha and theta range of the EEG after several antidepressants in an on-road experiment. Riemersma *et al.* (1977) found that heart rate decreased and heart rate variability increased in eight hours of night-time driving. The data from De Waard and Brookhuis (1991) confirmed this finding, however, after taking alcohol the effect was reversed (Figure 19.1). Initially heart rate was slightly decreased, but gradually increased as the BAC gradually decreased from 0.046 per cent to 0.035 per cent during the driving test.

De Waard and Brookhuis found a highly significant effect of alcohol (0.35–0.46 promille) on the amount of weaving (standard deviation lateral position), registering an increase of about 30 per cent. However, in the alcohol condition, steering wheel movements (both standard deviation and number of steering wheel reversals) were not affected. Petit *et al.* (1990), in a driving simulator study, reported considerable effects of alcohol on steering wheel movements; however, they did not calibrate the amount of alcohol to weight – their subjects were administered a whole bottle of wine.

Increases in energy in alpha- and theta-band of the EEG, indicative for decreases in subject's activation (Akerstedt and Thorsvall 1984; Akerstedt *et al.*, 1991) were found in an on-the-road experiment, coinciding with increases in amount of weaving (Brookhuis *et al.*, 1986). In this particular experiment, 6 out of 20 subjects were not able to finish their test ride in the condition that showed the highest effect on performance.

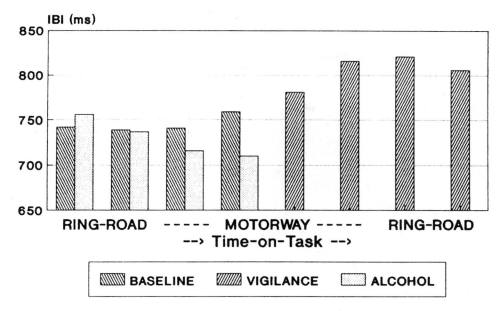

Figure 19.1 Inter-beat intervals (IBI) of the ECG while driving on a busy ring-road and quiet motorway, separated in to and from turning points, with and without alcohol.

19.3 Vigilance

Monotony is an important aspect of the driving task. The reason for this is that driving a car under relatively quiet conditions is relatively simple and monotonous for experienced drivers. Therefore, driving a car on a quiet highway is often considered to be a vigilance task (Wertheim, 1978), resulting from underload. The time-on-task related deactivation is making the driver accident-prone (Lisper *et al.*, 1986).

Underload will result in a deviation from the top of the inverted-U curve towards low activation and poor performance. This is nicely illustrated in two experiments that combine performance measures and physiology (Thorsvall and Akerstedt, 1987; Brookhuis *et al.*, 1986). After prolonged driving of a train or a car, driver's activation tends to diminish rapidly, as may be found by means of spectral analysis of the EEG. An increase in power in the theta- (4–8 Hz) and alpha-band (8–13 Hz) of the EEG-spectrum has been found to coincide with performance decrement while driving a car (Brookhuis *et al.*, 1986). Figure 19.2 illustrates this coincidence in high correlations of power in the alpha and theta band with the amount of weaving, measured as standard deviation of lateral position.

Petit *et al.* (1990) demonstrated a relationship between the handling of the steering wheel and the occurrence of alpha waves in the EEG. The state of vigilance, as indicated by the power in the alpha range, was found to correlate highly with specially developed steering wheel functions in most cases.

De Waard and Brookhuis (1991) also found effects on steering wheel behaviour with time-on-task (150 minutes of continuous driving). The standard deviation of the steering wheel movements increased and the number of steering wheel reversals per minute decreased, both highly significant. Subjects' activation, measured by a relative energy parameter ((alpha + theta)/beta), gradually diminishes from the start of the experiment. In Figure 19.3, both the relative EEG energy parameter and the

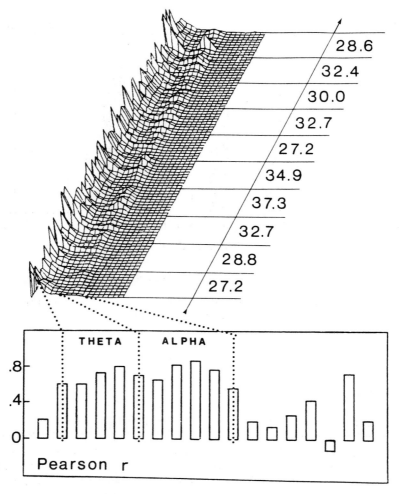

Figure 19.2 Top: EEG power density spectra (x-axis) for each 1-km segment in a 100-km driving test (y-axis) next to 10 standard deviations of lateral position in cm over 10 km each Bottom: barograph indicating product-moment correlations of the standard deviation lateral position and each 1 Hz band averaged across 10 km.

amount of weaving are depicted, showing that physiological signs of changes in driver state are readily followed by changes in driver behaviour.

Alcohol and medicinal drugs mostly aggravate these effects (Brookhuis et al., 1986), although this is not necessarily the case, depending on dose and type of drug.

Heart rate and heart rate variability can also serve as an index of mental load (Mulder, 1980). Time-on-task effects were found on performance, heart rate, heart rate variability and the 0.10 Hz component of the power spectrum of inter-beat-interval sequences (Coles, 1983; Brouwer and van Wolffelaar, 1985). Riemersma et al. (1977) found decreases in heart rate and increases in heart rate variability in eight hours of night-time driving. They concluded that the observed decreases of heart rate reflected mainly adaptation, including decreasing stress and habituation to sensory stimulation. The increases in heart rate variability are in line with this explanation.

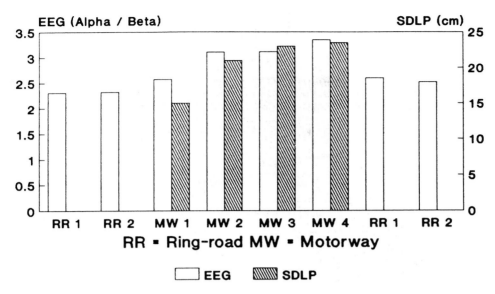

Figure 19.3 Average standard deviation lateral position (amount of weaving) and EEG ((alpha + theta)/beta) energy during 150 minutes of continuous driving

19.4 Task-load

Although the primary concern in driving performance research has been with mental underload and alcohol, high workload situations resulting in mental overload might well be a major cause in, for instance, intersection related accidents (Smiley and Smith, 1985; Smiley and Brookhuis, 1987). Overload emerges when two or more aspects of the driving task compete for attention (Michon, 1985), and this is often the case at intersections.

Secondary, i.e. subsidiary, tasks have been used to measure spare mental capacity while driving a motor vehicle (Brown *et al.*, 1969; Wetherell, 1981; Lisper *et al.*, 1986). The results of this type of measurement do not seem very promising. In so far as the experiments involved relevant aspects of the driving task situation, they indicated that various levels of workload, e.g. induced by levels of steering difficulty, can be demonstrated weakly by performance on the secondary task. Heart rate and heart rate variability might be more promising. In particular heart rate variability and the earlier mentioned 0.10 Hz component are reported to be reliable indices of increased workload (Egelund, 1985; Aasman *et al.*, 1987).

Brookhuis *et al.* (1991) found that heart rate increased when subjects in an on-the-road driving test carried out a subsidiary telephone task. In Figure 19.4 this effect is illustrated, showing effects of handling a telephone while driving, whereas training shows a considerable habituating effect in three weeks of daily practising.

Heart rate variability and its 0.10 Hz component showed a similar, systematic decrease in amplitude, indicating increasing workload. Effects of telephoning while driving were also visible in the standard deviation of the steering wheel movements, however, only on busier parts of the test track, not on the quiet motorway.

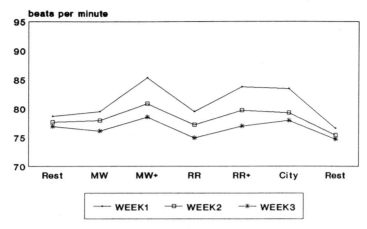

Figure 19.4 Heart rate (beats per minute), averaged across three weeks of practice, with (+) telephoning while driving and normal driving, separately for quiet motorway (MW), busier ring-road (RR) and city.

19.5 Discussion

Tonic changes

In the study of De Waard and Brookhuis (1991) it was found that tonic changes (visible in global ECG parameters) co-occurred with gradual changes in behavioural parameters, such as the standard deviation of the lateral position, which is in line with the findings of Riemersma *et al.* (1977) and Egelund (1985). In the experiments the effects of (potential) driver impairment in a vigilance situation are demonstrated in drivers' physiology and thereupon in safety-relevant driving parameters. Time-on-task affects driver state and consequently driver behaviour. Similarly, the effects of external factors (such as alcohol and drugs) are shown to be deleterious in many respects by many researchers.

Phasic changes

From several studies (Brookhuis *et al.*, 1986; Petit *et al.*, 1990) it appears that phasic changes (as reflected in EEG energy parameters) also precede demonstrable changes in behaviour, indicating that (even small) changes in driver state are followed by changes in relatively easily measured vehicle parameters such as steering wheel movements or lateral position control. Phasic measurements are important for the detection of driver impairment in the sense of overload, since in traffic overload usually arises fast.

In conclusion, combined measurements of driver's physiology and behaviour have demonstrated a relationship that could be used in the development phase of a driver impairment monitoring device on the basis of a limited number of unobtrusive vehicle parameters alone. In fact, in the European ATT (Advanced Transport Telematics) programme, as well as in the US IVHS (Intelligent Vehicle Highway Systems) programme such devices are under development.

References

Aasman, J., Mulder, G. and Mulder, L.J.M., 1987, Operator effort and the measurement of heart rate variability, *Human Factors*, **29**, 161–70.

Akerstedt, T. and Thorsvall, L., 1984, Continuous electrophysiological recordings, in Cullen, J.J. and Siegriest, J. (Eds), *Breakdown in Human Adaptation to Stress. Towards a multi-disciplinary approach, Vol. I*, The Hague: Martinus Nijhoff, pp. 567–84.

Akerstedt, T., Kecklund, G., Sigurdsson, K., Anderzén, I. and Gillberg, M., 1991, Methodological aspects on ambulatory monitoring of sleepiness, in *Proceedings of the workshop 'Psychophysiological Measures in Transport Operations'*, Köln: DLR, pp. 1–21.

Brookhuis, K.A. and de Waard, D., 1993, The use of psychophysiology to assess driver status, *Ergonomics*, **36**, 1099–110.

Brookhuis, K.A., Louwerens, J.W. and O'Hanlon, J.F., 1986, EEG energy-density spectra and driving performance under the influence of some antidepressant drugs, in O'Hanlon, J.F. and de Gier, J.J. (Eds), *Drugs and Driving*, London: Taylor & Francis, pp. 213–21.

Brookhuis, K.A., de Vries, G. and de Waard, D., 1991, The effects of mobile telephoning on driving performance, *Accident Analysis and Prevention*, **23**, 309–16.

Bookhuis, K.A., Volkerts, E.R. and O'Hanlon, J.F., 1987, The effects of some anxiolytics on car following in real traffic, in Noordzij, P.C. and Roszbach, R. (Eds), *Alcohol, Drugs and Traffic Safety*, T86, Amsterdam: Excerpta Medica, pp. 223–7.

Brouwer, W.H. and van Wolffelaar, P.C., 1985, Sustained attention and sustained effort after closed head injury: detection and 0.10 Hz heart rate variability in a low event rate vigilance task, *Cortex*, **21**, 111–19.

Brown, I.D., 1991, Prospects for improving road safety during the 1990s, in Singleton, W.T. and Dirkx, J. (Eds), *Ergonomics, health and safety*, Leuven: Leuven University Press, pp. 315–24.

Brown, I.D., Tickner, A.H. and Simmonds, D.C.V., 1969, Interference between concurrent tasks of driving and telephoning, *Journal of Applied Psychology*, **53**, 419–24.

Coles, M.G.H., 1983, Situational determinants and psychological significance of heart rate change, in Gale, A. and Edwards, J.A. (Eds), *Physiological Correlates of Human Behaviour, Vol. II: Attention and performance*, London: Academic Press, pp. 171–85.

de Waard, D. and Brookhuis, K.A., 1991, Assessing driver status: a demonstration experiment on the road, *Accident Analysis and Prevention*, **23**, 297–307.

Egelund, N., 1985, Heart rate and heart rate variability as indicators of driver work load in traffic situations, in Orlebeke, J.F., Mulder, G. and Vandoornen, L.J.P. (Eds), *Psychophysiology of cardiovascular control*, New York, London: Plenum Press, pp. 855–65.

Levine, J.M., Greenbaum, G.D. and Notkin, E.R., 1973, *The effects of Alcohol on Human Performance: A Classification and Integration of Research Findings*, American Institutes for Research.

Lisper, H.O., Laurell, H. and Vanloon, J., 1986, Relation between time to falling asleep behind the wheel on a closed track and changes in subsidiary reaction time during prolonged driving on a motorway, *Ergonomics*, **29**, 445–53.

Louwerens, J.W., Gloerich, A.B.M., de Vries, G., Brookhuis, K.A. and O'Hanlon, J.F., 1987, The relationship between drivers' blood alcohol concentration (BAC) and actual driving performance during high speed travel, in Noordzij, P.C. and Roszbach, R. (Eds), *Alcohol, Drugs and Traffic Safety – T86*, Amsterdam: Excerpta Medica, pp. 183–7.

Michon, J.A., 1985, A critical view of driver behaviour models: what do we know, what should we do?, in Evans, L. and Schwing, R.C. (Eds), *Human behaviour and traffic safety*, New York: Plenum Press, pp. 485–524.

Mulder, G., 1980, The heart of mental effort, PhD thesis, Groningen: Rijksuniversiteit Groningen.

Petit, C., Chaput, D., Tarrière, C., LeCoz, J.Y. and Planque, S., 1990, Research to prevent the driver from falling asleep behind the wheel, in *Proceedings of the 34th AAAM conference*, American Association of Automotive Medicine, pp. 505–23.

Riemersma, J.B.J., Sanders, A.F., Wildervanck, C. and Gaillard, A.W.K., 1977, Performance decrement during prolonged night driving, in Mackie, R.R. (Ed.), *Vigilance: Theory, operational performance, and physiological correlates*, New York: Plenum Press, pp. 41–58.

Seppala, T., Linnoila, M. and Mattila, M.J., 1979, Drugs, alcohol and driving, *Drugs*, **17**, 389–408.

Smiley, A. and Brookhuis, K.A., 1987, Alcohol, drugs and traffic safety, in Rothengatter, J.A. and de Bruin, R.A. (Eds), *Road Users and Traffic Safety*, Assen: Van Gorcum, pp. 83–105.

Smiley, A. and Smith, R.L., 1985, Driving task monitoring to reduce inattention-related accidents, Technical report, National Highway Traffic Safety Administration.

Thomas, D.B., Brookhuis, K.A., Muzet, A.G., Estève, D., Tarrière, C., Poilvert, C., Schrievers, G., Norin, F. and Wyon, D.P., 1989, Monitoring driver status: the state of the art, Technical report to the Commission of the European Communities, DRIVE Project V1004 (DREAM), Köln: TÜV-Rheinland.

Thorsvall, L. and Akerstedt, T., 1987, Sleepiness on the job: continuously measured EEG in train drivers, *Electroencephalography and Clinical Neurophysiology*, **66**, 502–11.

Wertheim, A.H., 1978, Explaining highway hypnosis: experimental evidence for the role of eye movements, *Accident Analysis and Prevention*, **10**, 111–29.

Wetherell, A., 1981, The efficacy of some auditory-vocal subsidiary tasks as measures of the mental load on male and female drivers, *Ergonomics*, **24**, 197–214.

Wiener, E.L., Curry, R.E. and Faustina, M.L., 1984, Vigilance and task load: in search of the inverted U, *Human Factors*, **26**, 215–22.

20 Fatigue and alcohol: interactive effects on human performance in driving-related tasks

David J. Mascord, Jeannie Walls and Graham A. Starmer

20.1 Introduction

Driver fatigue is now viewed as a major cause of serious traffic crashes in Australia, especially in the trucking industry (Haworth et al., 1989; Haworth, 1990). Camkin (1990) stated that fatigue may be a major factor in approximately 20 per cent of all road fatalities in New South Wales, increasing to 30 per cent in rural areas. In a questionnaire-based survey of truck drivers in Australia, 77.5 per cent of drivers rated fatigue as at least a substantial problem while 84.6 per cent of drivers reported experiencing fatigue, at least occasionally, while driving (Williamson et al., 1992).

The multifactorial nature of traffic crashes is often overlooked by both authorities and investigators (Corfitsen, 1986). While most studies attribute 40–55 per cent of highway fatalities solely to alcohol intoxication, it is likely that fatigue also contributes to a large proportion of these crashes (Ryder et al., 1981; Schwing, 1990). The involvement of alcohol may mask the presence of fatigue and the reverse may also be true. The interactions between alcohol and fatigue may be subtle. Alcohol can cause drowsiness at blood–alcohol concentrations (BACs) which are well below the *per se* limits for drivers (Landauer and Howat, 1983) and residual sedative effects may persist long after alcohol has been completely eliminated from the blood (Walsh et al., 1991).

Many of the manifestations of fatigue and alcohol intoxication are surprisingly similar. Ryder et al. (1981) found a consensus that both alcohol and fatigue reduced driver information-processing capacity and increased vehicle control variability. Complex decision-making tasks were more susceptible to alcohol, however, while simple tasks were more readily affected by fatigue.

Reaction speed has been found to decrease progressively over time both during prolonged driving (Lisper et al., 1971; 1973) and in performance on a vigilance task (Mascord and Heath, 1992). The effects of alcohol on reaction time have been found to depend on both the dose and complexity of the task (Moskowitz and Robinson, 1988). Simple reaction time is relatively resistant to the effects of alcohol but impairment of complex reaction time has been detected at BACs as low as 0.03 g/100 ml (Palva et al., 1982). Where fatigue and alcohol intoxication (0.13 g/100 ml) were present concomitantly, marked impairment of simple reaction time has been found (Corfitsen, 1982).

Tracking tasks are measures of perceptual-motor ability which are analogous to steering a vehicle (Moskowitz, 1973). An alert vigilant driver smoothly tracks the road geometry by making many fine steering adjustments. A fatigued driver, however, makes fewer fine adjustments and is forced to make more compensatory tracking responses. These coarse variations in lateral displacement often lead to an obvious meandering steering pattern (Seko et al., 1986). Increased steering wheel reversals and lateral position errors have been also been found under ethanol (Gawron and Ranney, 1988).

Pursuit tracking tasks, which require concomitant attention to two or more information sources, are sensitive to alcohol with substantial impairment occurring at BACs of 0.05 g/100 ml or even as low as 0.02 g/100 ml (Moskowitz et al., 1985).

Divided attention tasks are measures of the ability to attend to and process information coming from more than one source, selectively. The ability to divide attention efficiently is impaired by both fatigue (Mascord and Heath, 1992) and alcohol (Gruner, 1959; Vuchinich and Sobell, 1978). Impairment can be reliably detected at BACs of 0.05 g/100 ml (Moskowitz and Robinson, 1988) and has been reported with BACs as low as 0.015 g/100 ml (Moskowitz et al., 1985).

Alcohol and fatigue have a common site of action in the central nervous system (CNS), the reticular activating system (RAS). Alcohol has apparently biphasic effects on the CNS, but the primary effect is one of depression (Wallgren and Barry, 1970) causing sedation (MacLean and Cairns, 1982) and reducing attention (Moskowitz and Robinson, 1988). The RAS plays a critical role in maintaining cortical arousal and attention (Perrine, 1973) and is highly sensitive to low concentrations of alcohol (Maling, 1970). Complex behaviours, such as driving and divided attention task performance, depend on the integrative functions of the RAS for their precise execution (Perrine, 1973) and the sensitivity of these activities to alcohol impairment is believed to be mediated via the RAS.

Critical flicker fusion (CFF) frequency is a sensitive measure of CNS arousal (Curran, 1990) and correlates significantly with electro-encephalographic (EEG) data (Gortelmeyer and Wieman, 1982) and self-reports of alertness (Grandjean et al., 1977). In general, reduced alertness or sedation decreases the CFF threshold whereas an increase in alertness or arousal raises it. This is true for both natural and drug-induced effects (Hindmarch, 1982).

An inverse relationship has been demonstrated between heart rate (HR) and distance/hours driven (Lisper et al., 1973; Mackie and O'Hanlon, 1977; Fagerstrom and Lisper, 1977). HR, however, is not a sufficiently reliable indicator of driver fatigue because it is also sensitive to physical activity and traffic-induced stress, both of which will increase HR (Egelund, 1982).

An increase in heart rate variability (HRV) is considered to be a more sensitive indicator of driver fatigue because it is less susceptible to transient changes in HR. One approach is to use frequency domain spectral analysis of either the inter-beat interval (IBI) or HR which examines patterns of frequency oscillations in IBI/HR. These oscillations are thought to be determined by three different physiological control mechanisms which are associated with thermoregulation, mean arterial pressure (MAP) and respiration (Sayers, 1973). If the IBI/HR is subjected to time series Fourier analysis, a power spectrum containing three frequency component signals is obtained, which appear as three spectral peaks (Mulder and Mulder-Hajonides van der Meulen, 1973):

- the first signal peak (0.02–0.05 Hz) arises from the thermoregulatory vasomotor controlling system (Kitney, 1980);
- the second signal peak (0.08–0.15 Hz) from the baroreceptor system controlling MAP (Sayers, 1973); and
- the third is the result of respiratory sinus arrhythmia or changes in intramural pressure (Mulder and Mulder-Hajonides van der Meulen, 1973).

When a mental workload is imposed on a subject, the three spectral signals decrease in power, with the most prominent reduction occurring in the 0.1 Hz region. Thus, this region is considered to be the most sensitive to changes in mental alertness (Mulder and Mulder, 1981). In agreement, an increase in spectral power in 0.06–0.15 Hz frequency range has been noted during fatigue conditions (Mulder and Mulder, 1981). Egelund (1982) also found a significant linear relationship between HRV in the 0.05–0.15 Hz region and the distance driven. Neither respiratory rate and depth (Mulder and Mulder, 1981) nor light physical work (Hyndman and Gregory, 1975) affect the oscillations in the 0.1 Hz region. The HRV signal between 0.0–0.15 Hz is therefore regarded as a more sensitive indicator of fatigue than the standard deviation of HR. The effects of alcohol on HRV have received little attention but alterations in HRV, which may depend on both dose and tolerance, have been reported (Gonzalez Gonzalez et al., 1992).

Taken together, these findings suggest that more research is needed to establish effects of fatigue and alcohol, alone and in combination, on driving. The present study was designed to investigate impairment of human psychomotor skills related to driving induced by fatigue alone and by the combined effects of fatigue and alcohol. Age was considered to be an important variable in the study. Age and experience are highly correlated with crash-rate in that young, inexperienced drivers are involved in a disproportionately high number of crashes (Federal Office of Road Safety, 1991).

20.2 Methods

Subjects

Twenty subjects comprised the study group:

- ten were young, inexperienced drinkers and drivers (five male and five female, 21 years or less); and

- ten were older, experienced drinkers and drivers (five male and five female, 30–58 years).

The subjects were all in good health and not receiving any medication, apart from three female subjects who were stabilized on oral contraceptives. The objective of the study and the procedure to be followed were carefully explained to each subject and an informed signed consent was obtained. Ethical approval was obtained from the Ethics Review Committee of the Central Sydney Health Service.

Alcohol

The alcohol dose (0.62 g/kg body weight) was presented as vodka (37.5% v/v) mixed with an equal volume of orange juice. The placebo treatment was the same volume of orange juice with 5 ml of vodka floated on top for olfactory masking (Starmer and Bird, 1984).

Measurement of blood–alcohol concentrations (BACs) by breath analysis

Two methods were used.

- An Alcomat (Siemens, Germany) infra-red (3.4 μm) instrument was used to breath analyse subjects before and after the trial. In two field trials ($n = 466$ and 94), the correlation coefficients between BACs obtained on this instrument and by direct analysis of venous samples have been reported to be 0.977 and 0.998 respectively (Slemeyer, 1985).

- During the fatigue session, BACs were determined using a hand-held fuel cell device (Alco-Sensor III, Intoximeters Inc., St Louis, USA). The correlation coefficient between BAC readings on this instrument and by analysis of capillary blood is 0.92 (Starmer, 1985).

Body temperature and blood pressure

Mouth temperature was measured using a digital thermometer (MT-9001). Blood pressure was recorded using a digital sphygmomanometer (ALP K2 [05-115], Japan).

Fatigue session

The fatigue session was a 180-min three-way divided attention task consisting of a central pursuit tracking task, a secondary peripheral visual discrimination task and a random visual 'emergency' signal presented as a red light. The central pursuit tracking task and the secondary peripheral visual discrimination task were presented on an IBM 386 clone (Porro) computer with a black monochrome screen, a single disk drive and a keyboard fitted with a template to assist target selection. A steering wheel was used to operate a potentiometer which was connected to the games port of the computer.

The central pursuit tracking task consisted of two 15-mm vertical lines, 25 mm apart which moved irregularly across the screen from left to right. The subject was required to maintain the rectangular cursor (7 × 5 mm) within the boundaries of the two moving lines by operating the steering wheel. Each time the cursor left the boundaries of the two vertical lines, the subject was alerted by a tone signal and the time taken for the cursor to return within the boundaries (time off-target) was recorded. The program responded to the subjects' level of performance by adjusting the speed and irregularity of the moving display every 10 s according to the number of errors made by the subject. The mean line speed was monitored every 10 s as a delay and then averaged to give an indication of performance over the 180-min session – the higher the numerical value of delay, the poorer the performance.

The peripheral visual detection task consisted of four 15-mm circles, one in each corner of the screen. The circles were each bisected by a line which randomly flipped in its orientation every 10 s to one of four directions; horizontal, vertical, left oblique or right oblique. When one of the lines moved to the right oblique orientation (the designated target), the subject was required to respond as quickly as possible by pressing the corresponding key on the template key pad. A total of 60 targets were presented during the three-hour session, each one displayed for 20 s. The reaction time from the appearance of the designated target to the response was recorded. Errors, including both incorrect targets and missed targets, were also recorded.

A Toshiba laptop computer (model T1200) was used to generate a visual (red light) 'emergency' signal which appeared above the monitor (14 times per hour at random intervals). When the red light appeared (5 s display), the subject was required to press a foot pedal as quickly as possible. The reaction time was recorded.

Heart rate (inter-beat interval – IBI)

An infra-red monitor, connected to the Toshiba T1200 laptop computer was clipped to the subjects' ear lobe and used to record heart IBI. The heart beat from the ear lobe monitor was detected by a hardware interrupt, converted to a heart IBI and then averaged over 5 s by the computer.

Psychomotor test battery

The visual psychomotor test battery (Lemon, 1990) consisted of three tasks: divided attention, digit symbol coding (DSC) and visual analogue scales (VAS). The visual analogue scales probed for changes in perceived sedation, boredom, co-ordination and willingness to drive a motor vehicle. A Leeds Flicker Fusion Tester was used to measure critical flicker fusion (CFF) frequency (Curran, 1990).

Procedure

Each subject attended the laboratory on two occasions, at least five days apart, between 4.30 p.m. and 11.30 p.m. The subjects were asked not to consume any food for 3–4 hours prior to testing, to abstain from alcoholic beverages and to limit their caffeine intake to one cup of tea or coffee on the day of testing. They were also asked to try to ensure that they had adequate sleep on the night before testing. On arrival at the laboratory, the subjects were breath tested to ensure that they were alcohol-free. They then completed a questionnaire detailing age, gender and drinking and driving history, and signed a consent form. They were then assigned to the active or placebo treatment according to a randomized schedule. Recordings of blood pressure and temperature were then taken while the subject was seated.

On the first occasion, the subject was familiarized with the fatigue session task (10 minutes) and the psychomotor test battery. The subject then completed a pre-test run on the psychomotor test battery and CFF. The placebo or alcoholic drink was then consumed evenly over a 20-minute period under close supervision. Subjects then carried out the three-hour fatigue session.

On completion of the fatigue session, subjects once again carried out the psychomotor and CFF tests and recordings of blood pressure and temperature were made. The subjects were offered a meal at this point and were then driven home.

20.3 Results

The divided attention and heart rate IBI data were collected over a continuous three-hour session. The data were reduced into three separate one-hour segments and examined for differences between the segments. The mean scores for each time segment are shown in Table 20.1. All variables were analysed using a repeated measures analysis of variance. The within-subject factors were treatment (placebo, alcohol) and time-on-task (1, 2, 3 hours). The between-subject factor was age (young, old). The Newman–Keuls test statistic was used for *post-hoc* comparisons between means.

Blood–alcohol concentrations (BAC)

The BACs were measured after hours 1, 2 and 3. The BACs attained by the younger subjects were 0.064, 0.049 and 0.033 g/100 ml and those by the older subjects were 0.059, 0.046 and 0.031 g/100 ml respectively.

Table 20.1 Mean (±s.e.m.) changes in divided attention and heart rate variables over time

Variable	Time on task (hours)			Significance level (p) for the effect of:		
	1	2	3	time	treatment	age
Tracking control (delay)	63.43 (13.2)	76.95 (17.7)	83.18 (19.5)	0.000	n.s	n.s
Tracking (time off target)	243 (42.21)	322 (56.78)	353 (73.95)	0.000	0.008	0.05
Peripheral target detection (msec)	3189 (332)	3231 (323)	3330 (380)	n.s	n.s	n.s
Peripheral target detection (errors)	9.0 (2.31)	10.3 (1.84)	9.9 (2.34)	n.s	0.006	n.s
Reaction time to red light (msec)	954 (112)	964 (126)	908 (141)	n.s	0.002	n.s
Heart Rate (HR)	80.1 (4.01)	80.7 (3.96)	77.9 (4.46)	0.010	0.001	n.s
HR variability (HRV) (sd)	6.02 (0.99)	6.22 (0.76)	6.31 (0.68)	n.s	0.048	0.02
HR spectral power (0.02–0.04 Hz)	2.22 (0.15)	2.20 (0.14)	2.20 (0.13)	n.s	0.002	0.02
HR spectral power (0.08–0.15 Hz)	1.90 (0.11)	2.05 (0.12)	2.04 (0.12)	0.001	0.001	0.05
HR spectral power (0.25–0.35 Hz)	1.64 (0.14)	1.83 (0.11)	1.88 (0.13)	0.000	n.s	n.s

n.s = $p > 0.05$.

Divided attention task (tracking control)

A highly significant ($F_{2,32} = 15.59$; $p < 0.001$) reduction in tracking control over time was found. Subjects were found to be significantly worse in the second hour ($q_{2,32} = 4.73$; $p < 0.05$) and third hour ($q_{3,32} = 12.33$; $p < 0.05$) than in the first hour. This effect was mainly due to a highly significant ($F_{2,32} = 12.33$; $p < 0.001$) interaction of age with time, which revealed that older subjects were more likely to be impaired in the second ($q_{3,32} = 4.73$; $p < 0.05$) and third hours ($q_{4,32} = 7.35$; $p < 0.05$) while the younger subjects were able to maintain a reasonably constant tracking control over time. No significant difference in tracking control between the young and older subjects was found during the first hour ($q_{3,32} = 0.45$; $p > 0.05$).

Tracking (time-off-target)

The amount of time spent off-target in the tracking task was found to significantly increase with time-on-task ($F_{2,32} = 21.29$; $p < 0.001$). Again, this effect was mainly due to a significant age with time interaction ($F_{2,32} = 7.52$; $p < 0.01$). Older subjects were found to be significantly impaired during the second ($q_{3,32} = 4.89$; $p < 0.05$) and third hours ($q_{4,32} = 7.18$; $p < 0.05$) when compared with their performance during the first hour. The amount of time spent off-target by the younger subjects was found to increase over time, although there was no significant difference between the first and third hours ($q_{4,32} = 1.71$; $p > 0.05$). The performance of the young and older subjects was similar during the first hour ($q_{2,32} = 1.06$; $p > 0.05$). The amount of time spent off-target was significantly increased by alcohol ($F_{1,16} = 9.15$; $p < 0.01$). The treatment by age interaction term was not significant, however ($F_{1,16} = 1.41$; $p > 0.05$).

Peripheral target detection
No significant differences were detected for age, treatment or time-on-task for the reaction time measure of the peripheral detection task. It can be seen from Table 20.1, however, that there was a general increase in reaction time over the three-hour period. Alcohol was found to increase significantly the number of errors made by all subjects when selecting the peripheral targets ($F_{1, 16} = 9.57; p < 0.01$). The greatest number of errors were made in the second hour of the experiment by subjects in both treatment groups.

Reaction to the 'emergency' light
Alcohol was found to increase significantly the time taken to respond to the 'emergency' light ($F_{1, 16} = 13.43; p < 0.01$). A significant age by time interaction was found ($F_{2, 32} = 3.29; p = 0.049$) although *post-hoc* comparisons between the treatment means failed to find any significant differences. The younger subjects were found to improve in their response to the 'emergency' light over time while the older subjects were slower to respond in the final two hours compared to their response rate in the first hour of the experiment.

Heart rate inter-beat interval (IBI)
The individual heart rate IBIs (msec) were recorded, averaged every 5 s and saved to disk every hour. This resulted in 720 IBIs per hour. The IBIs were then transformed (60 000/IBI) into individual heart rates. Spectral decomposition using BMDP1T frequency domain spectral analysis was then carried out on the individual heart rates after subtraction of the sample mean and linear trend removal (Bloomfield, 1976). Heart rate (HR), heart rate variability (s.d. of HR) and the average log values of the power contributions for the frequency bands of 0.02–0.04 Hz (temperature), 0.08–0.15 Hz (blood pressure) and 0.25–0.35 Hz (respiration) were analysed using an analysis of variance with repeated measures.

Heart rate
A significant reduction in mean heart rate was found during the three-hour session ($F_{2, 32} = 5.28; p < 0.01$). The heart rate was significantly lower during the last hour of the session compared with both the first ($q_{2, 32} = 3.55; p < 0.05$) and the second hour ($q_{3, 32} = 4.52; p < 0.05$). Alcohol administration caused a highly significant increase in mean heart rate ($F_{1, 16} = 16.61; p < 0.001$). The older subjects were found to have a lower mean heart rate than the younger subjects but the difference failed to reach significance ($p = 0.075$).

Heart rate variability
The standard deviation of the mean heart rate was used to examine variability in heart rate over time but the differences were not significant ($F_{2, 32} = 0.37; p > 0.05$). Younger subjects were found to have greater heart rate variability than the older subjects ($F_{1, 16} = 7.39; p < 0.05$). Alcohol caused a significantly greater overall increase in heart rate variability than placebo ($F_{1, 16} = 4.47; p < 0.05$). A significant time by treatment by age interaction was found ($F_{2, 32} = 4.59; p < 0.05$). Heart rate

variability increased over time after the placebo treatment in younger subjects but decreased after the alcohol treatment. In contrast, the heart rate variability increased over time in the older subjects after both the placebo and alcohol treatments.

Heart rate spectral power (0.02–0.04 Hz)
The temperature component of the power spectra did not alter over time, although there was a highly significant increase in the log power value after alcohol compared to placebo ($F_{1,16} = 13.37; p < 0.01$). Older subjects were found to have significantly smaller log power values than the younger subjects after both the alcohol and placebo treatments ($F_{1,16} = 7.14; p < 0.05$).

Heart rate spectral power (0.08–0.15 Hz)
The blood pressure component of the power spectra was found to increase significantly over time ($F_{2,32} = 9.29; p < 0.001$). The mean log power values for the second and third hours were found to be significantly greater than that for the first hour ($q_{2,32} = 3.88; p < 0.05$) and ($q_{3,32} = 3.85; p < 0.05$) respectively. A significant interaction between age and time was also found ($F_{2,32} = 3.29; p < 0.05$). Older subjects were found to have significantly larger log power values in the second and third hours on task compared to the first hour ($q_{2,32} = 4.02; p < 0.05$) and ($q_{3,32} = 4.40; p < 0.05$) respectively. Although the younger subjects were found to have increased log power values in the second and third hours of the session, the values were not significantly different from the first hour. A significant increase in the log power spectra for the blood pressure component was also found when subjects received alcohol as opposed to placebo ($F_{1,16} = 14.79; p < 0.01$). The increases over time after alcohol were 1.97, 2.15 and 2.12 Hz and after placebo, 1.82, 1.94, and 1.97 Hz.

Heart rate spectral power (0.25–0.35 Hz)
A highly significant increase over time was found in the log power spectra for respiration ($F_{2,32} = 12.22; p < 0.001$). The differences between the first hour and both the second ($q_{2,32} = 3.65; p < 0.05$) and third hour ($q_{3,32} = 4.62; p < 0.05$) reached significance. There was a trend ($p = 0.078$) for the younger subjects to have larger log power values than the older subjects.

Measurements made before and after the fatigue session
The data were analysed using a repeated measures analysis of variance. The within-subject factors were time (pre-, post-), and treatment (placebo, alcohol). The between subjects factor was age (young, old). *Post-hoc* comparisons were carried out using the Newman–Keuls test statistic.

Divided attention task
A significant decrease in tracking control was found at the end of the fatigue session ($F_{1,16} = 7.53; p < 0.05$) when compared with pre-session performance. No other significant effects of age or treatment were identified. No significant differences were found

for age, treatment or time of testing for the peripheral detection reaction time component of the divided attention task.

Digit symbol coding

A significant age difference in reaction time was found ($F_{1, 16} = 14.73; p < 0.01$). Younger subjects were found to respond more rapidly to the appearance of the symbols than their older counterparts. A significant interaction between age, treatment and time was also found ($F_{1, 16} = 5.50; p < 0.05$). Older subjects were significantly more impaired after the fatigue session when given alcohol than when they received the placebo treatment ($q_{4, 16} = 4.49; p < 0.05$).

Critical flicker fusion (CFF) frequency

The three-hour fatigue session resulted in a highly significant reduction of CFF threshold ($F_{1, 16} = 15.17; p < 0.01$). Post-hoc analysis of the significant interaction ($F_{1, 16} = 12.89; p < 0.01$) between age, treatment and time revealed that, after alcohol, the CFF threshold of older subjects was reduced to a significantly greater extent after the fatigue session than that of younger subjects ($q_{4, 16} = 9.43; p < 0.05$). The CFF thresholds were also lower at the end of the fatigue session in both age groups in the placebo condition but the differences from pre-session values just failed to reach significance.

Oral temperature

A significant treatment by time interaction was found ($F_{1, 16} = 8.64; p < 0.01$). A significant post-test decrease in mouth temperature was found when the subjects received the placebo beverage ($q_{4, 16} = 7.02; p < 0.05$) but not after alcohol. A significant age by time interaction was also found ($F_{1, 16} = 4.77; p < 0.05$) and although *post-hoc* analysis failed to reveal significant differences, there was a post-test decrease in oral temperature in the older subjects which was absent in the younger subjects.

Blood pressure

The mean systolic and diastolic blood pressures were all within the normal range and were not significantly influenced by age, time of testing or treatment.

Subjective effects – visual analogue scales

No significant effects were attributable to age. A highly significant ($F_{1, 16} = 55.75; p < 0.001$) increase in sedation and a highly significant ($F_{1, 16} = 85.88; p < 0.001$) reduction in co-ordination over time were reported by all subjects after the fatigue session. All subjects reported a significant increase in boredom by the end of the experiment ($F_{1,16} = 8.97; p < 0.05$), although this effect was not as strong as those of sedation and reduced co-ordination. All subjects reported a significant ($F_{1, 16} = 26.27; p < 0.001$) reduction in their willingness to drive a motor vehicle by the end of the experimental session. No significant separation between the effects of alcohol or placebo over time was found.

20.4 Discussion

This study aimed primarily to explore the effects of fatigue on driving-related skills in the laboratory and to investigate the influence of alcohol on the development of fatigue. A three-way divided attention task was used to monitor the performance of subjects for a continuous three-hour period. To provide additional information on the effects of prolonged performance on the task and to correlate the findings with previous experience in the laboratory (Yap et al., 1993), standard tests (two-way divided attention, digit symbol coding, critical flicker fusion (CFF) frequency and visual analogue scales) were carried out by the subjects before and after the fatigue session. Several physiological parameters (heart rate, systolic and diastolic pressure and temperature) were also monitored.

Measurements made during the fatigue session

The majority of the subjects rated peripheral detection as the most difficult element of the task and indicated a tendency to concentrate on this component while progressively ignoring the others, especially tracking. Detection of the emergency light was rated as the easiest of the three components.

The highly significant reduction in tracking performance observed in this study over time is consistent with the results of most other studies which have examined fatigue in a laboratory setting. For example, Dureman and Boden (1972) noted an increased frequency of tracking (steering) errors over four hours of continuous operation of a driving simulator which correlated highly with other performance decrements, such as reaction time and the ability to efficiently divide attention. Mascord and Heath (1992) found a sharp deterioration in tracking performance in the final 35 min of a 140-min vigilance task, which was interpreted as an indicator of the onset of a fatigue-induced reduction of operator performance. The tracking performance of the older subjects was found to be significantly worse than that of the younger subjects, both in terms of control and time spent off-target. However, although the younger subjects were able to maintain a relatively constant tracking control throughout the session, they also spent progressively more time off-target, which is probably an indicator of fatigue.

Under alcohol, there was a significant increase in the time spent off-target, compared with placebo, and there was a strong trend towards a greater reduction of tracking performance in the older subjects. Moskowitz and Burns (1971) have also reported a deleterious interaction between age and alcohol with a resultant additive detrimental effect on the rate of information processing. These findings are in apparent conflict with those of a number of other studies (Lisper et al., 1971; Fagerstrom and Lisper, 1977; Peacock, 1991) which have found an association between youth and greater impairment of driving skills, especially after alcohol ingestion. An explanation may lie in the overlearned nature of the driving task as opposed to laboratory measures. When young (mean age of 18) and older (mean age of 42) subjects were required to carry out a novel laboratory task, there were no differences in the performance of the two groups under either alcohol or placebo conditions (Peacock, 1991).

Reaction speed and error rate in the peripheral detection task did not increase significantly over time. This finding, although unexpected, can probably be explained in terms of a trade-off effect, considering the extensive deterioration in tracking control which occurred concomitantly. Trade-off typically involves a progressive impairment of the ability to attend to more than one incoming source of information as fatigue sets in. Alcohol significantly increased the number of errors in the peripheral detection component of the task. This finding is consistent with the well-known ability of alcohol to reduce the capacity to divide attention efficiently (Moskowitz and Robinson, 1988).

Alcohol significantly impaired the speed at which subjects reacted to the emergency light. This finding is consistent with previous research which has shown that BACs below 0.05 g/100 ml are associated with significant impairment of emergency braking performance (Laurell, 1977). There was an age-related difference in this measure in that whereas the younger subjects generally improved their performance over the three-hour session, that of the older subjects deteriorated. The reported simplicity of detection of the emergency light may explain why no time-on-task effects were noted. This finding is unexpected because alcohol has been shown to increase simple reaction time progressively as increasing time is spent on a vigilance task (Gustafson, 1986). A key factor in determining the extent of performance degradation, especially that of reaction time, is subject motivation (Stave, 1977). At the end of the fatigue session the younger subjects reported using the emergency light as an arousing challenge to counteract the monotony of the task and this may explain the observed improvement in the performance of these subjects over time.

A decrease in heart rate has been found both after prolonged periods of driving (Mackie and O'Hanlon, 1977; Fagerstrom and Lisper, 1977) and in extended performance on a laboratory vigilance task (Mascord and Heath, 1992). A consensus view is that a reduction in heart rate tends to occur when the task difficulty and information-processing demands are low (McDonald and Cameron, 1974) and heart rate is increased where high information-processing demands are imposed (Perelli, 1980).

The slowing of heart rate over time, found in this and previous studies, has been consistently correlated with decreased driver alertness and increased fatigue (Dureman and Boden, 1972). The decline in heart rate is considered to be the result of an increase in parasympathetic vagal tone arising from changes in CNS arousal levels. Significantly reduced production of adrenaline and noradrenaline after extended periods of driving, as indicated by decreased urinary excretion of metabolites, suggests that a reduction in sympathetic tone may also be involved (Mackie and O'Hanlon, 1977).

In this study, a highly significant increase in heart rate occurred after alcohol. The younger subjects were found to have a faster heart rate than the older subjects. This is consistent with previous findings which have shown that acute doses of alcohol elevate heart rate (Nelson et al., 1979; Martin et al., 1985). Alcohol and acetaldehyde are known to facilitate the release of catecholamines from the adrenal medulla (Pohorecky and Brick, 1988) and basal levels of arousal may also have been increased, which is consistent with the disinhibitory effects of low doses of alcohol (Perrine, 1973). The lower mean heart rates found in the older subjects may simply reflect age differences since the heart rate tends to decline with age (Brandfonbrener et al., 1955). There is no apparent explanation for the lack of effect of alcohol on heart rate in these subjects.

An increase in heart rate variability (HRV – measured as the standard deviation of heart rate) has often been encountered under boring monotonous conditions and it has been suggested that this indicates a reduction in CNS arousal and a diminished readiness to respond to unpredictable events (O'Hanlon and Kelly, 1977). An increased HRV has also been demonstrated as a function of driving hours and fatigue (O'Hanlon, 1973) and a decreased HRV has been shown to be associated with increasing information-processing load (Perelli, 1980). While the majority of investigators have concluded that HRV is a useful indicator of general fatigue levels (Mackie and O'Hanlon, 1977; Riemersma et al., 1977), some findings have been inconsistent and, here, the investigators have regarded HRV as a poor predictor of fatigue (Sayers, 1973; Egelund, 1982).

No studies have yet investigated the combined effects of alcohol and fatigue on HRV. In this study the differences in HRV were not significant over time. Overall, alcohol was found to increase HRV and this may simply reflect alcohol-induced fluctuations in the autonomic control of heart rate.

The signal power of the temperature component of the power spectra did not alter over time. This was expected because this low frequency band has not been found to correlate with changes in cognitive processing (Mulder and Mulder, 1981). A highly significant increase in the power signal was observed after alcohol, however. This is consistent with the results of a previous study which found a power increase in the temperature oscillation (0.02–0.06 Hz), 20–30 minutes after an acute dose (0.3 g/kg) of alcohol (Gonzalez Gonzalez et al., 1992). It is known that alcohol interferes with thermoregulatory mechanisms and may also lower the hypothalamic set-point for temperature control (Pohorecky and Brick, 1988). Thus, the change in the power of the temperature component observed in this study after alcohol would be expected because this frequency oscillation arises from thermoregulatory mechanisms (Kitney, 1980).

The signal power of the cardiac frequency oscillations associated with blood pressure controlling mechanisms (0.08–0.15 Hz) has been found to be extremely sensitive to changes in cognitive processing (Mulder and Mulder, 1981) and an increase in power has been found in fatigued states (Egelund, 1982; Mascord and Heath, 1992). Changes in this frequency band are thus held to provide a reliable indicator of fatigue. In this study, the blood pressure component of the power spectra was found to increase significantly over time, indicating an increase in fatigue. This is in accord with previous research by Egelund (1982) where an increase in the spectral power in the 0.05–0.12 Hz frequency band after a prolonged on-road driving task was reported, and by Mascord and Heath (1992), who found that the development of fatigue after a 140-minute vigilance task corresponded with an increase in spectral power in the 0.08–0.12 Hz region. The older subjects were found to be more fatigued than the younger subjects, as indicated by the significantly larger spectral power values in the second and third hours compared to the first hour of the fatigue session. In contrast, the younger subjects were much less fatigued and the increases in spectral values in the second and third hours were not significantly different from those in the first hour. Alcohol treatment was found to induce significantly greater increases in the spectral values than placebo. In contrast to the results presented here, acute doses of alcohol (0.3 g/kg) have been reported to decrease the power of the 0.08–0.15 Hz frequency oscillation but only during the period 20–60 minutes after dosage (Gonzalez Gonzalez et al., 1992). The finding in this study therefore may indicate that the effects of fatigue were much more prominent than those of alcohol on the frequency oscillations in this band over time.

A highly significant increase in spectral power of the frequency oscillations associated with sinus arrhythmia (0.25–0.35 Hz region) was found with increasing time spent on the fatigue session task. This is consistent with previous research which has found decreases in the power of these respiratory frequencies to be associated with both laboratory and industrial tasks which impose high mental workloads (Sayers, 1973). Spectral power values of the respiratory frequency band have also been found to increase with the onset of fatigue (Mascord and Heath, 1992). Alcohol has been shown to decrease power at this frequency during the first hour after dosage (Gonzalez Gonzalez et al., 1992) but such an effect was not seen in this study, possibly indicating that the fatigue effect was dominant.

The decrements in performance found in this study are indicative of the development of fatigue over time and are in accord with the consensus opinion that performance declines as a function of time in vigilance and divided attention tasks (Dureman and Boden, 1972; Mackie and O'Hanlon, 1977; Mascord and Heath, 1992). These decrements were associated with a decrease in heart rate and an increase in the spectral power of the blood pressure and respiration frequency regions, which are regarded as highly sensitive and reliable indices of fatigue.

Measurements made before and after the fatigue session

Driving is an example of a skilled divided attention task (Moskowitz, 1973) and laboratory tasks which require time-sharing ability have been found to be valid and reliable predictors of the effects of fatigue and low doses of alcohol (peak BACs > 0.05 g/100 ml) on driving performance (Moskowitz and Robinson, 1988). In this study, although there was a significant reduction in tracking control after the fatigue session, no interactive effects with alcohol were evident. The latter finding probably reflects the decline of the BACs to a low level (approximately 0.03 g/100 ml) by the end of the session and the development of acute tolerance (Franks *et al.*, 1977).

Digit symbol coding is a choice reaction time task which requires continuous information processing and sustained attention on a highly repetitive task and is well known to be highly sensitive to alcohol, with reliably identifiable deficits at BACs as low as 0.03–0.04 g/100 ml (Moskowitz and Robinson, 1988). In this study, the older subjects reacted more slowly than the younger subjects, whether they received alcohol or not. The older subjects were also significantly more affected by the combination of alcohol and fatigue, a finding which conflicts with that of Peeke *et al.* (1980) who reported that the effects of alcohol and sleep deprivation on choice reaction time (and most of their other measures) were antagonistic.

Critical flicker fusion (CFF) frequency was used as an index of overall central nervous system arousal which measures the ability of the central nervous system to process distinct pieces of sensory information (Curran, 1990). The significant lowering of the CFF threshold of the whole group after the session confirms the sensitivity of this measure to reduced levels of arousal/alertness which are a consequence of fatigue (Hindmarch, 1982). Older subjects were also found to be significantly more affected than younger subjects after alcohol. This difference is unlikely to be due to the effects of alcohol alone because repeatable and consistent reductions in CFF have been reported only at BACs of 0.07 g/100 ml. Below this level, the findings are controversial (Starmer, 1989) and it should be noted that the BACs at the end of the fatigue session were approximately 0.03 g/100 ml.

Reports on the relationship between body temperature and fatigue have been few and inconsistent. Thackery *et al.* (1977) found a decrease in oral temperature after the completion of a one-hour vigilance task which correlated well with the high boredom and fatigue of the group. Others (Richardson *et al.*, 1982) found increased sleepiness to be associated with an almost significantly decreased oral temperature and Mackie and O'Hanlon (1977) found no change in oral temperature after a 600 km highway drive, even though reduced levels of vigilance and performance decrements associated with fatigue were clearly evident. A fall in body temperature is most likely to occur after exposure to a cold environment (Sherwood, 1989) but in the present study, the room was heated and a more likely explanation for the significant decrease in mouth temperature after the fatigue session lies in the prolonged period of virtual immobility (Nelson, 1985).

Lowered arousal levels have often been considered to be associated with a reduction in sympathetic nervous system activity with consequent peripheral vasodilatation and result in a decrease in blood pressure (Berne and Levy, 1988). However, in this study, blood pressure was found not to be affected by either time-on-task or alcohol.

After completion of the fatigue session under both placebo and alcohol conditions, all subjects reported that they felt considerably more bored and fatigued and were less willing to drive a vehicle. The symptoms of fatigue were reflected as highly significantly increased perceptions of sedation and reduced co-ordination. These results are in accord with those of many studies, including that of Fuller (1984) who found that truck drivers at the end of their shift experienced increased feelings of drowsiness, lack of energy and had strong urges to stop driving. Similar feelings of intense drowsiness and boredom were reported by subjects after completion of a four-hour simulator study which correlated well with performance decrements (Dureman and Boden, 1972).

From the results presented in this study, it has been shown that older drivers may be more susceptible to the effects of fatigue than younger drivers and that this impairment is potentiated by low doses of alcohol. Since the effects of fatigue found in the older subjects may be attributed to a generalized age-related decline in co-ordinative and attentional capacity, the finding of alcohol-related impairment in the older subjects was unexpected. The older subjects reported, on average, a drinking history of 15–20 years while the younger subjects reported a drinking history of only a few years. Although the mean amount of alcohol consumed per week was similar for both older (10.9 g) and younger (11.2 g) subjects, it would be expected that the older subjects may have developed a greater degree of tolerance to alcohol and would therefore exhibit less impairment from alcohol than the younger subjects.

References

Berne, R.M. and Levy, N.L., 1988, Regulation of the heartbeat, in Berne, R.M. and Levy, N.L. (Eds), *Physiology*, St Louis: C.V. Mosby Company, pp. 451–71.
Bloomfield, P., 1976, *Fourier Analysis of Time Series: An Introduction*, New York: John Wiley & Sons.
Brandfonbrener, M., Landdowne, M. and Shock, N.W., 1955, Changes in cardiac output with age, *Circulation*, **12**, 557.
Camkin, H., 1990, Introduction: Fatigue and road safety, in *Workshop on Driver Fatigue: Report of Proceedings*, Sydney: Roads and Traffic Authority Road Safety Bureau, CR 2/90.
Corfitsen, M.T., 1982, Increased viso-motoric reaction time of young tired drunk drivers, *Forensic Sci. Int.*, **20**, 121–5.
Corfitsen, M.T., 1986, Fatigue in single car fatal accidents, *Forensic Sci. Int.*, **30**, 3–9.
Curran, S., 1990, Critical flicker fusion techniques in psychopharmacology, in Hindmarch, I. and Stonier, P.D. (Eds), *Human Psychopharmacology, Measures and Methods, Vol. 3*, Chichester: John Wiley & Sons, pp. 21–38.
Dureman, E. and Boden, C.H., 1972, Fatigue in simulated car driving, *Ergonomics*, **15**, 299–308.
Egelund, N., 1982, Spectral analysis of heart rate variability as an indicator of driver fatigue, *Ergonomics*, **25**, 663–72.
Fagerstrom, K.O. and Lisper, H.O., 1977, Effects of listening to car radio, experience, and personality of the driver on subsidiary reaction time and heart rate in a long term driving task, in Mackie, R.R. (Ed.) *Vigilance*, New York: Plenum Press, pp. 73–85.
Federal Office of Road Safety, 1991, *Road Safety 2000: Strategic Plan*, Canberra: Federal Department of Transport, pp. 8–11.
Franks, H.M., Starmer, G.A. and Teo, R.K.C., 1977, Acute adaptation of human cognitive and motor functions to ethanol, *Proc. 7th Int. Conference on Alcohol, Drugs and Traffic Safety*, Melbourne, Australia, 190–9.
Fuller, R.G.C., 1984, Prolonged driving in convoy: The truck driver's experience, *Acc. Anal. & Prev.*, **16**, 371–82.

Gawron, V.J. and Ranney, T.A., 1988, The effects of alcohol dosing on driving performance on a closed-course and in a driving simulator, *Ergonomics*, **31**, 1219–44.

Gonzalez Gonzalez, J., Mendez Llorens, A., Mendez Novoa, A. and Cordero Valeriano, J.J., 1992, Effect of acute alcohol ingestion on short term heart rate fluctuations, *J. Stud. Alcohol*, **53**, 86–90.

Gortelmeyer, R. and Wieman, H., 1982, Retest reliability and construct validity of critical flicker fusion, *Pharmacopsychiatr.*, **15**(S1), 24–8.

Grandjean, E., Baschera, P., Martin, E. and Weber, A., 1977, The effects of various conditions on subjective states and critical flicker fusion, in Mackie, R.R. (Ed.) *Vigilance*, New York: Plenum Press, pp. 331–9.

Gruner, O., 1959, Constitutional differences in the effect of alcohol, *Dtsch. Z. Gesamte Gerichtl. Med.*, **49**, 84.

Gustafson, R., 1986, Effect of moderate doses of alcohol on simple auditory reaction time in a vigilance setting, *Percept. Motor Skills*, **62**, 683–90.

Haworth, N.L., 1990, Driver fatigue and road accidents, in *Proc. Seminar on New Methods to Reduce the Road Toll*, Melbourne: Royal Australasian College of Surgeons.

Haworth, N.L., Heffernan, C.J. and Horne, E.J., 1989, Fatigue in truck accidents, Melbourne: Monash University Accident Research Centre, Report No. 3.

Hindmarch, I., 1982, Critical flicker fusion frequency (CFFF): The effects of psychotropic compounds. *Pharmacopsychiatr.*, **15**, 44–8.

Hyndman, B. and Gregory, J.R., 1975, Spectral analysis of sinus arrythmia during mental loading, *Ergonomics*, **18**, 225–70.

Kitney, R.I., 1980, An analysis of the thermoregulatory influences on heart-rate variability, in Kitney, R.I. and Rompelman, O. (Eds), *The Study of Heart Rate Variability*, New York: Oxford University Press, pp. 81–107.

Landauer, A.A. and Howat, P., 1983, Low and moderate alcohol doses, psychomotor performance and perceived drowsiness, *Ergonomics*, **26**, 647–57.

Laurell, H., 1977, Effects of small doses of alcohol on driver performance in emergency traffic situations, *Accid. Anal. & Prev.*, **9**, 191–201.

Lemon, J., 1990, *The Rozelle Test Battery. A Computerised Testing Instrument for Visumotor and Cognitive Performance*, Sydney: National Drug and Alcohol Research Centre, Technical report No: 9, pp. 1–14.

Lisper, H.O., Laurell, H. and Stening, G., 1973, Effects of experience of the driver on heart rate, respiration rate, and subsidiary reaction time in a three hours continuous driving task, *Ergonomics*, **16**, 501–6.

Lisper, H.O., Dureman, I., Ericsson, S. and Karlsson, N.G., 1971, Effects of sleep deprivation and prolonged driving on subsiduary auditory reaction time, *Acc. Anal. & Prev.*, **2**, 335–41.

Mackie, R.R. and O'Hanlon, J.F., 1977, A study of the combined effects of extended driving and heat stress on driver arousal and performance, in Mackie, R.R. (Ed.) *Vigilance*, New York: Plenum Press, pp. 537–58.

Maclean, A.W. and Cairns, J., 1982, Dose-response effects of ethanol on the sleep of young men, *J. Stud. Alcohol*, **43**, 434–43.

Maling, H.M., 1970, Toxicology of single doses of ethyl alcohol, in *International Encyclopedia of Pharmacology and Therapeutics*, New York: Pergamon Press, pp. 110–17.

Martin, N.G., Oakeshott, J.G., Gibson, J.B., Starmer, G.A., Perl, J. and Wilks, A.V., 1985, A twin study of psychomotor and physiological responses to an acute dose of alcohol, *Behaviour Genetics*, **15**, 305–47.

Mascord, D. and Heath, R.A., 1992, Behavioural and psychological indices of fatigue in a visual tracking task, *J. Safety Res*, **23**, 19–25.

McDonald, W., and Cameron, C., 1974, The use of behavioural methods to access traffic hazards, *Australian Road Research*, **14**, 1–27.

Moskowitz, H., 1973, Laboratory studies of the effects of alcohol on some variables related to driving, *J. Safety Res.*, **5**, 185–9.

Moskowitz, H. and Burns, M., 1971, Effect of alcohol on the psychological refractory period, *Q. J. Stud. Alcohol*, **32**, 782–90.

Moskowitz, H. and Robinson, C.D., 1988, *Effects of Low Doses of Alcohol on Driving-related Skills: A Review of the Evidence*, Washington, D.C.: National Highway and Traffic Safety Administration, Report No. DOT HS 807 280, pp. 1–97.

Moskowitz, H., Burns, M.M. and Williams, A.F., 1985, Skills performance at low blood alcohol concentrations, *J. Stud. Alcohol*, **46**, 482–5.
Mulder, G. and Mulder-Hajonides van der Meulen, W.R.E.H., 1973, Mental load and the measurement of heart rate variability, *Ergonomics*, **16**, 69–84.
Mulder, G. and Mulder, L.J.M., 1981, Information processing and cardiovascular control, *Psychophysiology*, **18**, 392–401.
Nelson, R.N., 1985, Accidental hypothermia, in Nelson, R.N., Rund, D.A. and Keller, M.D. (Eds), *Environmental Emergencies*, Philadelphia: W.B. Saunders Co., pp. 1–8.
Nelson, T.M., Laden, C.J. and Carlson, D., 1979, Perceptions of fatigue as related to alcohol ingestion, *Int. J. Waking & Sleeping*, **3**, 115–35.
O'Hanlon, J.F., 1973, *Heart Rate Variability: A New Index of Driver Alertness/fatigue*, Human Factors Research, Inc. Society of Automotive Engineers, pp. 1–8.
O'Hanlon, J.F. and Kelly, G.R., 1977, Comparison of performance and physiological changes between drivers who perform well and poorly during prolonged vehicular operation, in Mackie, R.R. (Ed.) *Vigilance*, New York: Plenum Press, pp. 87–109.
Palva, E.S., Linnoila, M., Routledge, P. and Seppala, T., 1982, Actions and interaction of diazepam and alcohol on psychomotor skills in young and middle aged subjects, *Acta Pharmacologica et Toxicologica*, **54**, 528–34.
Peacock, M., 1991, The effects of low doses of ethanol on human psychomotor performance and car driving ability, Thesis, Department of Pharmacology, The University of Sydney.
Peeke, S.C., Callaway, E., Jones, R.T., Stone, G.C. and Doyle, J., 1980, Combined effects of alcohol and sleep deprivation in normal young adults, *Psychopharmacology*, **67**, 279–87.
Perelli, L.P., 1980, Effects of fatigue stressors on flying performance information processing, subjective fatigue, and physiological cost indices during simulated long duration flight, *Dissertation Abstracts International*, **44**, 391–2.
Perrine, M.W., 1973, Alcohol influences on driving-related behaviour, *J. Safety Res.*, **5**, 165–84.
Pohorecky, L.A. and Brick, J., 1988, Pharmacology of ethanol, *Pharmacol. Ther.*, **36**, 335–427.
Richardson, G.S., Carskadon, M.A., Orav, E.J. and Dement, W.C., 1982, Circadian variation of sleep tendency in elderly and young adult subjects, *Sleep*, **5**, S82–94.
Riemersma, J.B.J., Sanders, A.F., Wildervank, C. and Gaillard, A.W., 1977, Performance decrement during prolonged driving, in Mackie, R.R. (Ed.) *Vigilance*, New York: Plenum Press, pp. 41–58.
Ryder, J.M., Malin, S.A. and Kinsley, C.H., 1981, *The Effects of Fatigue and Alcohol on Highway Safety*, Washington, DC: National Highway Traffic Safety Administration, DOT-HS-805-854.
Sayers, B., 1973, Analysis of heart rate variability. *Ergonomics*, **16**, 17–32.
Schwing, R., 1990, Exposure controlled highway fatality rates: Temporal patterns compared to some explanatory variables, *Alcohol, Drugs & Driving*, **5** and **6**, 275–85.
Seko, Y., Kataoko, S. and Senoo, T., 1986, Analysis of driving behaviour under a reduced state of alertness, *Int. J. Vehicle Design*, **7**, 318–30.
Sherwood, L., 1989, *Human Physiology: From Cells to Systems*, New York: West Publishing Company, pp. 296 and 609–21.
Slemeyer, A., 1985, Quantitative breath analysis with the Alcomat, *Atemalkohol*, **1**, 105–15.
Starmer, G.A., 1985, *Alcohol Interlock Devices and Self-test Devices, Report Series 7/85*, Adelaide: South Australian Department of Transport, Division of Road Safety, pp. 1–81.
Starmer, G.A., 1989, Effects of low to moderate doses of ethanol on human driving related performance, in Batt, R.D. and Crow, K.E. (Eds.) *Human Metabolism of Alcohol, Vol. 1, Pharmacokinetics, Medicolegal Aspects and General Interests*, New York: CRC Press, pp. 101–30.
Starmer, G.A. and Bird, K.D., 1984, Investigating drug-ethanol interactions, *Br. J. Clin. Pharmac.*, **18**, 2S–35S.
Stave, A.M., 1977, The effects of cockpit environment on long-term pilot performance, *Human Factors*, **19**, 503–14.
Thackery, R.I., Bailey, J.P. and Touchstone, M.R., 1977, Physiological, subjective and performance correlates of reported boredom and monotony while performing a simulated radar control task, in Mackie, R.R. (Ed.) *Vigilance*, New York: Plenum Press, pp. 203–13.
Vuchinich, R.E. and Sobell, M.S., 1978, Empirical separation of physiologic and expected effects of alcohol on complex perceptual motor performance, *Psychopharmacology*, **60**, 81–5.
Wallgren, H. and Barry, H., 1970, Ethanol in the metabolism, in Wallgren, H. and Barry, H. (Eds),

Actions of Alcohol: Volume 1, Biochemical, Physiological, and Psychological Aspects, Amsterdam: Elsevier, pp. 125–8.

Walsh, J.K., Humm, T., Muelbach, M.J., Sugerman, J.L. and Schweitzer, P.K., 1991, Sedative effects of ethanol at night, *J. Stud. Alcohol*, **52**, 597–600.

Williamson, A.M., Feyer, A., Coumarelos, C. and Jenkins, T., 1992, *Strategies to Combat Fatigue in the Long Distance Road Transport Industry. Stage 1: The Industry Perspective*, Canberra: Federal Office of Road Safety, Report No. CR 108, pp. 1–225.

Yap, M., Mascord, D.J., Starmer, G.A. and Whitfield, J.B., 1993, Studies on the chronopharmacology of ethanol, *Alcohol and Alcoholism*, **28**, 17–24.

21 Developing a high fidelity ground vehicle simulator for human factors studies

James W. Stoner

21.1 Introduction

This chapter describes the Iowa Driving Simulator (IDS) facility at the University of Iowa and examines the requirements for high-fidelity driving simulation. It also establishes a baseline for future simulator development by providing a description of simulators currently operating at the IDS facility.

Several of the facility's high-fidelity simulation research projects are examples of other simulator research being funded in the US. These projects demonstrate that research simulators have fundamentally different objectives than training devices. Multiple simulation scenarios and releases require database development and system integration support. Simulation control systems must be flexible to ensure that new databases support unanticipated demands. The system architecture must be based on the general principles of modular software so that the system is able to support evolving requirements.

21.2 Requirements for high-fidelity driving simulation

Since the 1930s, technical simulator development has focused on traditional flight training applications. It has not been until the last ten years, that interactive ground vehicle simulation and training have been developed. Because of this recent development and the small number of research simulators in operation, the market for software and integration services has been limited. As it stands, ground vehicle simulators adapt cueing system hardware originally designed for flight simulators (e.g. motions bases, image generators and host computers). A real-time driving simulator provides images to the driver's eyes, audio cues to the ears, and motion and vibration to the ears and vestibular system.

While flight training certification requirements determine visual system display specifications, such as luminance, contrast, and resolution, driving simulation imposes an entirely new set of requirements. These requirements include adequate visual resolution to satisfy appropriate detection and recognition distances for traffic control systems. High resolution depth cues are critical for many driving manoeuvres. High visual update and refresh rates are also necessary to display close proximity moving models. The display configuration in terms of viewports or channels is quite different from flight or fixed wing aircraft.

Roadway geometries must be correct with respect to engineering roadway specifications. Vertical and horizontal geometries, as well as cross-sectional elements must be absolutely correct with respect to engineering standards. Accurate representation of super-elevation, spiral transitions, and tangent runout on horizontal highway curves requires multiple levels of detail and a large number of textured polygons.

Evaluation of the terrain and road surface representation by the vehicle dynamics code must occur within the contact patch of each of the four tyres, so a correlated road surface database with variable mesh size is one solution. All simulation subsystems require a knowledge of the precise road geometry to ensure absolute correlation.

Scenario control subsystems impose a large computational load on the host computer and the large number of moving models is beyond the capacity of current image generator architectures. Scene authoring must be flexible and able to accommodate a wide variety of experimental requirements. Hundreds of smart vehicles and pedestrians must be modelled not only as visual objects, but as smart interacting entities with varying behaviours and responses. Moving models must have both scripted and random behaviours. All simulations must have a deterministic replay capability.

It is for all these reasons, that the capabilities of the current generation of high-fidelity image generators will be severely stressed by ground vehicle applications. Vendors must manufacture computer image generation (CIG) systems that support more levels of detail, greater texture mapping capability, more model select features, and increased dynamic coordinate system control.

Audio cues must accurately present three-dimensional sound sources to represent tyre, power train, wind and traffic noises. Audio correlation with the other simulation subsystems allows both full-directional location of the cue and front-to-rear and side-to-side dynamic panning with Doppler shift.

Motion cue and performance specifications for flight simulation are not directly transferable to ground vehicle applications. Empirical research is limited because only two currently operating simulators dedicated to driving research have six degree of freedom motion systems. Translational platforms, such as those used in NASA's vertical motion simulator, allow for a much larger envelope of motion, but were built to support helicopter training and evaluation. The Daimler Benz facility in Berlin is, however, mounting a hexapod on a track providing realistic lateral or longitudinal acceleration cues to the driver.

All four tyres need to sample the road surface for roughness and z-value at very high update rates. The update rate on an image generator is inadequate to provide the necessary information required by vehicle dynamics for realistic vehicle response. A correlated surface mesh with high-frequency interrogation capability and varying levels of resolution is used to provide the road surface data at the required frequencies.

Vehicle dynamics run in real time and are based on multi-body, basic-principles models which can simulate limit handling conditions. Integration rates of 150–300 Hz assure numerical stability of the dynamics calculations. Accurate calculation of steering rack forces is necessary for realistic control loading that allows the driver to consider the vehicle to be under his or her control. Approximation of the geometry and measurements of the steering linkage can result in unnatural handling.

Tyre models require high rates of integration, which are only achievable through multi-rate algorithms. Anti-lock braking system (ABS) and automatic stability response (ASR) applications cannot be simulated without accurate representation of tyre-spin dynamics. Simple point-follower models will not allow high-frequency features such as curbs, potholes, and paving joints.

Direct coupling of the control loading systems (i.e. steering, acceleration and braking) to vehicle dynamics accurately represents of torsional forces and vibration to be reflected to the driver through the vehicle control elements. An accurate steering system computer model takes into account the non-linearities and compliance attributes of system mechanics and hydraulics. Software has only recently been developed at Iowa to model power steering, brakes and power train in real time.

Motion cue masking and the nature of false cues during ground vehicle simulation are very different from flight. Adaptive washout algorithms and motion control software tuned for driving simulation have great promise, but are only now being developed. The ability to represent the high-frequency motion associated with pavement slab joints, curbs, potholes, wind buffeting, and other special effects are necessary for sustained driver realism. Correlating the motion cues with visuals, vehicle dynamics and audio is also necessary.

21.3 Description of the Iowa driving simulation (IDS) facility

The IDS research staff at the Center for Computer-Aided Design (CCAD) provides support for a variety of human factors and vehicle performance studies. The range of simulators available at the centre include:

1. a hexapod motion platform with changeable cabs and 300° projection dome;
2. a fixed-base simulator with a 220° screen display; and
3. generic driving cabs with IBM and SGI workstation graphics systems with three-dimensional imaging for construction and agricultural equipment experiments.

Figure 21.1 is a photograph of the open dome on the motion system simulator with a military vehicle traversing a virtual proving ground. Figure 21.2 is a view from behind the driver in the Ford Taurus cab during an actual simulation experiment.

Two four-channel textured graphics computer image generation systems are available and can be used in any of the simulation bays. A parallel super-computer achieves real-time calculation of vehicle dynamics using models that retain an engineering design level of fidelity. A modular software environment provides flexibility in adapting to a myriad of research environments encountered in virtual prototyping, training, and research applications for both on-road and off-road vehicles. All four simulation bases use pin-for-pin compatible instrumentation connections for data collection and can be configured to use the appropriate visual, vehicle dynamics, host computer, and scenario control, systems.

Integration of the IDS system required an evaluation of various system hardware components of the system (visual system, motion system, vehicle dynamics), to determine an appropriate operating environment, and an evaluation of each subsystem for interface specification, computational, memory, and latency requirements. A four-processor Harris Nighthawk 5808 parallel computer serves as the host computer, using a real-time operating system that allows deterministic periodic scheduling. An Alliant FX2800 computer, with 26 high-speed RISC I860 based processors, provides the power to perform real-time dynamics and complex scenario control. High-speed data interfaces connect the Harris host computer with the Alliant and the host computer for the image generators, which are themselves very specialized computers. Figure 21.3 is a schematic diagram of the IDS software and hardware systems.

A simulator operating system written by the University of Iowa permits a well-defined and expandable interface between subsystems, supporting both subsystem intercommunication and synchronization. This development process allowed simultaneous development of the simulation control program and the associated subsystem software

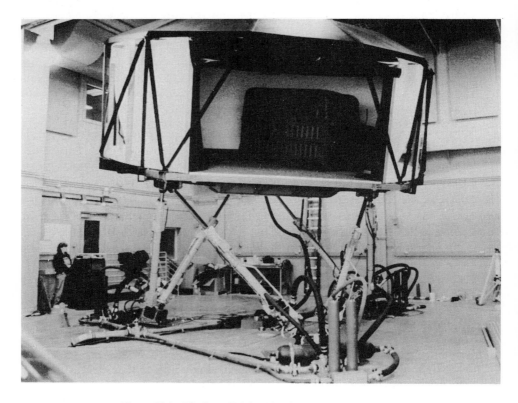

Figure 21.1 The Iowa Driving Simulator (IDS) motion base.

for visuals, control loading, scenario control, vehicle dynamics, audio, data collection and instrumentation. Interface modules were installed in the Harris computer for analogue and digital communication with the instrumented vehicle cab and with the motion system.

The motion base used in the IDS is a Singer-Link Heavy Payload system, capable of generating motion with six degrees of freedom. The response characteristics of this motion system are as follows.

- Maximum payload: 10 872 kg (24 000 lb)
- Linear acceleration: 1.1 g
- Linear travel: ±1.14 m (45 in)
- Angular acceleration: 200°/sec^2
- Angular travel: ±30°
- Frequency response: 8 Hz

The motion control algorithm reproduces motion cues that the driver would normally feel in the vehicle, while avoiding motion commands outside the envelope of the motion base. The motion control system uses both classical washout algorithms and adaptive

Developing a high fidelity ground vehicle simulator for human studies 211

Figure 21.2 The view from the interior of a Taurus cab.

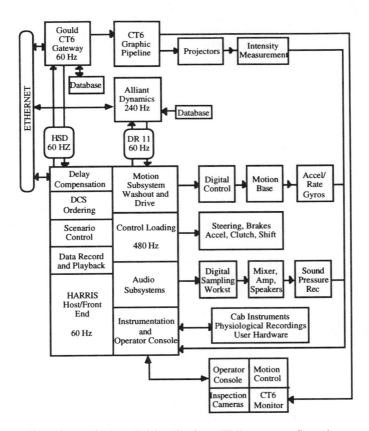

Figure 21.3 The Iowa Driving Simulator (IDS) system configuration.

algorithms. Chassis linear accelerations and angular rates are high-pass filtered to maintain high-frequency onset cues required for vehicle control, while eliminating low-frequency large amplitude motions that produce sustained accelerations. Tilt co-ordination provides co-ordinated low frequency cues lost during filtering. Integration of filtered accelerations and angular rates obtains the chassis orientation of the simulator cab, which is then transformed into the coordinates of the individual actuators on the motion system.

Both the Ford Taurus and High Mobility Multi-purpose Wheeled Vehicle (HMMWV) cabs have simulation controlled instruments including speedometer, tachometer, fuel and temperature gauges, and indicator lights. Potentiometers on the brake, accelerator and gear shift and a digital encoder on the steering wheel provide control position information. The potentiometer values are sent to A/D converters on the Harris host computer. The control loading model of the steering system uses the forces on the tie-rod ends calculated by the vehicle dynamics subsystem, translated to an appropriate voltage. Torque feedback to the driver's hands comes from a DC servo disk motor connected to the steering column.

Either an Evans and Sutherland CT-6 or ESIG 2000 system produce the high-fidelity computer images. Visual database development and modelling tools and visual data management software support both image generation devices. Real-time images are rendered, based on eye point, moving model, and environmental information from the simulator's scenario control and vehicle dynamics systems. These very capable systems provide up to four configurable channels of high-resolution textured graphics. The system can support simultaneous high scene density and smoothness of motion. Both systems allow for multiple levels of detail for database models, and provide smooth fading transitions between levels. A wide range of atmospheric effects, including variable sun-angle, illumination, fog and lightning, contribute to the induced realism.

A maximum of 256 dynamic coordinate systems support moving models, and individual or chained moving elements. Single-body vehicle models or fully articulated models can be presented. Each dynamic coordinate system manipulates up to 128 different moving models. Sequencing of models allows the animation of special effects. The IDS vehicle library currently provides representations of 125 unique automobile, truck, bus and motorcycle models. Instancing of models allows their presentation multiple times within a single database for simulation of interactive traffic. A fully animated walking pedestrian model is also available.

The display system placed on the motion base uses 5.5 m (18 ft) diameter dome segments and four hardened high performance multi-sync projectors mounted in the dome structure. A cross-fire configuration using three of the projectors provides a forward field-of-view (FOV) of 200° horizontal by 40° vertical. A fourth projector creates either a rear view or a segment of interest view on the appropriate dome segments. This approach provides high-quality graphics, as well as adequate mechanical rigidity to withstand accelerations and forces imposed by the motion system.

Databases can be map-correlatable to real-world locations, or be entirely fictional. Among the specialized database development software already obtained and implemented to meet IDS visual database needs are an advanced procedural modelling system from Evans and Sutherland, a workstation-based interactive modelling package from Software Systems (called MultiGen), and a number of custom modelling algorithms developed at The University of Iowa for automating construction of particular database features.

An IDS modelling package constructs realistic textured roadway models for driving databases. It facilitates the creation of complex roadways, including constructions involving parabolic vertical curves and banked horizontal curves with variable rates of super elevation. Specific modelling code for providing various centre stripe combinations, roadside markers, street lights, shoulders and ditches, and a full library of traffic signs and control devices support the basic road surface modelling routines. Software System's MultiGen provides automatic terrain model generation, based on digital topographical information from either the US Geological Survey or the Defense Mapping Agency. It also includes interactive texture creation, editing, and application tools for both synthetic and photo-derived texture maps.

The audio systems produce the auditory cues that are necessary for realism in the driving environment, including engine, wind, road and passing car sounds. The audio system uses off-the-shelf, MIDI-based digital sampling systems. The digital approach permits the use of source audio data recorded from the real vehicle environment. Microphones placed in an actual operating vehicle record each sound type individually, ensuring accurate reproduction in the simulator. A series of digital recordings, using digital analogue tape to reduce signal losses, cover the entire envelope of operation of the vehicle. Each sound type is digitally sampled at several discrete operating points. During actual simulations, evaluation of vehicle dynamic and environment state variables produce representative sounds from dithered and blended sound samples.

The audio system consists of a 16-voice digital-sampling workstation, a MIDI controlled mixer, and eight speakers placed around the vehicle cab on the simulator platform. The system, using the mixer, supports cross channel fading and phasing of sounds to give sound directionality. The sampler is capable of reproducing 16-bit audio samples at 44 K Hz, producing an output bandwidth of 20–20 K Hz with a dynamic range of over 95 dB.

The IDS vehicle dynamics subsystem receives inputs from the operator through the vehicle controls. It then predicts the vehicle motion and generates output for IDS visual, motion, audio and scenario control subsystems.

The vehicle equations of motion consist of differential and algebraic equations (DAE) derived from multi-body dynamics formulations. Models of vehicle specific

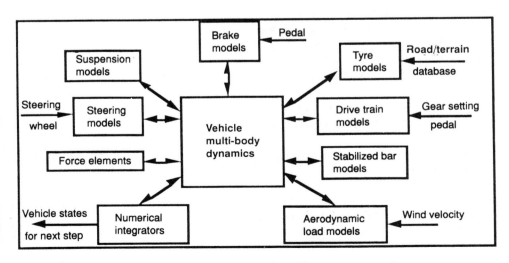

Figure 21.4 The vehicle modelling structure.

subsystems such as steering, brake, power train and aerodynamics interact with the general multi-body dynamics to predict the motion of the vehicle. Figure 21.4 shows the relationship of dynamics with other subsystems. Outbound arrows communicate vehicle state information and returning arrows illustrate forces acting on the vehicle.

The scenario control software residing on the host computer, manages all controllable features or elements within the simulator databases. This definition includes control of simulated traffic, dynamic traffic control devices, environmental conditions (weather and illumination), animated special effects and any other dynamic or transient features.

Scenario control creates pseudo-random (i.e. random, but repeatable) distributions of autonomous vehicles that interact and behave in a near-natural fashion, having awareness of each other and of the simulator subject's vehicle, as well as an understanding of the rules of the road. Simultaneous management of over 100 vehicles, including the introduction and removal of vehicles from activity, permits high-density traffic in a simulation release. Each vehicle may be preprogrammed to specify behavioural characteristics, including the preferred speeds and headways as well as gap acceptance tendencies. All moving objects are controlled by position and orientation commands passed to the image generation system for use by the appropriate dynamic coordinate systems and their associated models. Each vehicle model also has control provided for proper dynamic display of its brake lights and turn signals.

In addition to general traffic simulation, the system allows for the programming of specific preplanned vehicular or object behaviour, to meet custom experimental design requirements. Such behaviour might include, among a myriad of possible examples, a pedestrian or other unexpected obstacles coming into the path of the operator's vehicle under certain triggering conditions. Programmed traffic control devices operate on either fixed-time intervals, or variable in either semi-actuated or full-actuated mode depending on current traffic arrival rates or densities. Control standard semaphores, turn lights, flashing red or yellow lights, or even rail crossing control devices become elements of the simulated environment. Environmental effects managed by the scenario control system include sun position and illumination level, visibility and ground fog effects, sky colour and cloudiness, wind direction and strength, and lightning.

The system models the behaviour of various objects within a given scenario, as defined by a scenario file and a number of road database files. The scenario file provides experiment specific definitions and controls. Road database files composed of space–curve lane and surface definitions based on global roadway specification, exactly match the roadway models of the visual database.

21.4 Description of current projects

Current research projects being performed at the Iowa Driving Simulation Facility fall into the following categories:

- virtual prototyping of vehicles,
- virtual prototyping of infrastructure and controls systems,
- human performance studies,
- driving behaviour and fitness to drive, and

- basic simulator technology development.

The Advanced Research Projects Agency (ARPA) land systems office h a major project at the Center for Computer-Aided Design to develop an in-the-loop virtual prototyping capability for ground vehicles and to demonstrate its use as an integral part of the ground vehicle acquisition process. A demonstration project based on a virtual prototype quality simulator for a high mobility multipurpose wheeled vehicle (HMMWV), using a sophisticated vehicle dynamics model coupled with a complete four-wheel drive power-train model, producing an accurate representation of the complex vehicle's performance. A HMMWV cab has been fully instrumented and outfitted with control loading hardware, then mounted in the visual dome.

The high-fidelity simulation required the creation of a detailed visual database of the Munson and Churchville test courses at the Aberdeen Proving Grounds using MultiGen and proprietary database packages created at the University of Iowa and Evans and Sutherland. This computer database accurately captures the geometry, appearance and surface characteristics of the test course and its surrounding cultural features. An associated high-resolution terrain database is fully correlated with the visual database, providing spatial resolution as accurate as a few inches where required to represent terrain or road surface features for use by the vehicle dynamics subsystem. Interrogation of the correlated databases in real time by the simulation allows the software to locate precisely each wheel of the simulated vehicle on the terrain surface at each integration time step. Vehicle and driver performance under different loading and environmental conditions will be compared with empirical results. The virtual prototyping facility will be used for a multi-year simulation-based design, after the initial validation period. The objectives of this programme are to develop linkages to distributed simulation networks and to show the qualitative impact on the ground vehicle acquisition process.

The TravTek Camera Car Evaluation project involves a safety and use ability evaluation of the TravTek Advanced Traveler Information System (ATIS) based on on-road field studies of systems installed in rental vehicles in the Orlando, Florida market. The TravTek system consists of both a moving map navigation display and yellow page's interrogation. This human factors study is being conducted to assess driving task intrusion, navigation performance, unsafe driving behaviour, experience, age and other factors affecting operator use of the system.

An IDS capabilities study, funded by the Federal Highway Administration, assessed the overall simulation performance of the Iowa Driving Simulator. A high-fidelity correlated database developed specifically for the study fed the visual, audio and vehicle dynamics subsystems. The experimental portion of the study focuses on a range of drivers performing navigation, hazard avoidance, vehicle control and navigation tasks under both motion and fixed-base conditions. The specific objectives of this research are to:

- assess the IDS capabilities in several performance dimensions,

- provide data that can be statistically analysed by the FHWA to identify vehicle manoeuvres that are particularly difficult and/or unsafe for drivers to perform, and

- develop the first of a series of standardized driving simulator scenarios.

The CCAD is a subcontractor to Honeywell, Inc. on a four-year project to evaluate the human factors aspects of automated highway design. A series of experiments conducted on the IDS provides a tool for evaluating driver performance in simulated intelligent vehicle/highway system (IVHS) environments, including computer-controlled headway maintenance in designated lanes under a variety of traffic flow, speed and average headway conditions. Drivers are evaluated with respect to their ability to perform the required traffic control functions to include entering and exiting the platoon, collision avoidance and route navigation.

The ATIS and commercial vehicle operations (CVO) components of the IVHS project is being performed under subcontract to Battelle to develop human factors guidelines for addressing the use of ATIS. Investigations of various ATIS prototypes require a variety of levels of vehicle simulation. Initial experiments will use personal computer-based simulation systems with projected monitor graphics and a generic vehicle cab. A number of ATIS prototypes will be evaluated and final systems designed and tested. The high-fidelity simulation capability of the IDS will be used in the final phases of this project to test fully the human factors interfaces to these systems.

The University of Iowa will be providing support services to develop performance specifications for rear-end collision avoidance systems. The various warning system devices will be tested in a variety of simulated environments during the actual design phase.

A project addressing scene-authoring requirements for driving simulation research funded by the National Highway Traffic Safety Administration supports the development of software allowing assembly of simulation scenes by an experimenter through a point and click interface. The elements of the database required for the specific experiments are divided into tiles containing various roadway geometries, intersection components and terrain. Macro tiles and super tiles are defined by combining sets of basic tiles. A corresponding scenario control database exists for each tile, allowing traffic at different densities to be generated as specified by the experimenter. Experimenters can create preprogrammed scripted models of vehicles or other objects as part of their release.

A project evaluating alternative technologies for prevention of single vehicles leaving roadway accidents is scheduled to begin in the summer of 1994. Simulation of high accident propensity conditions provided a test bed for warning devices.

21.5 National advanced driving simulator (NADS)

The US Congress has authorized the US Department of Transportation (USDOT) to construct a NADS to support research relating to driving safety and vehicle and highway design. The NADS is intended to provide capabilities well beyond those of any other existing, or planned, ground vehicle simulator in the world. The NADS facility requires a large motion platform with a high-resolution graphics display dome. The motion felt by the driver is produced by both the large actuator systems and small high-frequency vibrators mounted on the motion platform. Graphics computers and display devices render 360° × 35° field of view, high-resolution scenes to be viewed by the driver. Audio feedback to the driver is provided by a computer-controlled audio system.

The USDOT contracted with the National Science Foundation to conduct a competitive solicitation to recommend a host university site for the NADS. The National Science Foundation competition was performed in conformance with the process used to select engineering research centres. At the conclusion of the review process, the National Science Foundation blue-ribbon panel unanimously recommended the University of Iowa as the NADS host site. On 20 January 1992, the Department of Transportation officially awarded the NADS to the University of Iowa. Contracts have been awarded to two consortia to develop initial concepts and preliminary designs during a one-year design off. A single team will then be selected to implement the selected concept. The NADS will become operational at The University of Iowa in 1998.

Research has been initiated on NADS design issues to establish system requirements and parameters. Simulation-based studies must be performed to determine what motion system parameters are required.

22 The development of a driver alertness monitoring system

John H. Richardson

22.1 Introduction

The European PROMETHEUS programme was launched in 1986 to encourage pre-competitive collaborative research within the automotive industry. The objectives of the programme are to develop and apply new technologies in order to increase traffic safety and road system capacity, and to reduce vehicle emissions. Its aims are therefore consistent with other intelligent vehicle highway systems (IVHS) programmes.

In the programme's 'Safe Driving' stream, research effort is being invested in areas such as vision enhancement, collision avoidance and proper vehicle operation. It is within the last area that the Ford Motor Company (UK) and the HUSAT Research Institute are undertaking research on driver alertness monitoring.

Driver fatigue is now being accepted as a major contributory factor to road accidents despite the considerable difficulty in quantifying the problem (O'Hanlon, 1978). Although Harris (1977), for example, found a clear relationship between time spent driving and likelihood of accident involvement, other investigators such as Hamelin (1987) have found stronger effects for time of day and sleep pattern disruption. This latter effect is consistent with typical findings regarding feelings of drowsiness and unintentional sleep episodes during the day. These show pronounced peaks between 2 a.m. and 7 a.m. and, to a lesser extent, between 2 p.m. and 5 p.m. (Carsakadon et al., 1985). Experimental investigations of driver fatigue have proved surprisingly contradictory, despite a large body of literature. There is certainly no simple, causal, relationship between time spent driving, fatigue and driving impairment.

The current project was conceived to provide a diagnostic tool that could be used to investigate driver fatigue in field research and, if successful, form the basis for a reliable in-vehicle warning system. The potential benefits provided by such a system are increased if current developments in IVHS technologies gain widespread implementation.

Many of the intelligent in-vehicle systems currently under development could result in increased demands in terms of driver vigilance – either through increased information presentation (e.g. navigational support and medium range pre-information) or through the need to monitor automated functions (e.g. intelligent cruise control) which were previously under the driver's direct control. In both cases, the drivers' ability to respond appropriately will depend on their maintenance of a sufficient level of arousal. However, the problem of maintaining high levels of attention while an individual is undertaking a prolonged, monotonous and undemanding activity is well established.

In order to ensure that a driver is capable of responding when required – either to a normally occurring, but infrequent, driving demand, a critical incident or to a new automated 'support' system – it is essential that the driver's own status is monitored and a warning given whenever this falls below the necessary level. It is also conceivable that output from a driver status monitor could be used directly to adapt the parameters employed in other intelligent subsystems. For example, the minimum headway for an intelligent cruise control system might be increased if the driver was judged to be under aroused.

A long-term perspective is being taken to the development of the system with the initial goal being a fatigue warning or advisory system. The development of active support systems for the driver are not envisaged until the end of the decade and systems which actively intervene in the vehicle's control are not expected to be available until early in the next century. Several years' basic research has already been spent on the development of this device and it is envisaged that, given the complexity of the issues involved, a considerable future effort will also be necessary.

22.2 Basic strategy

In the phased development programme, the initial stage required the direct measurement of driver fatigue, along with measures of driver performance, during extended driving sessions. In the subsequent 'training' stage, a neural net would be trained off-line to identify patterns in the driver performance datasets which were consistent with the behaviour of tired drivers – as indicated by the direct measures of fatigue. In the final evaluation stage, the net would be given driver performance data in real time during an extended drive and be expected to predict driver state.

The original conception of the system involved continuous psycho-physiological driver monitoring based on the measurement of eye-blink 'events'. A video camera mounted on the test vehicle's dashboard would capture an image comprising the driver's head and immediate surrounds. Image-processing software operating in real time would then identify those portions of the captured image corresponding to the driver's eyes and track them. Changes in the image's contrast level in the eye region could then be used to determine whether the eyes were open or closed on a frame-by-frame basis and thus derive an estimate of eye-blink frequency. This approach enjoyed the advantage of being continuous and completely unobtrusive to the driver.

In conjunction with the psycho-physiological driver measurement, an instrumented vehicle would be developed with autonomous data recording capabilities. Quantitative data relating to the driver's actual performance are relatively easy to gather from contemporary motor vehicles. Sensor systems around the vehicle already monitor inputs such as throttle, brake, speed and steering to control anti-lock brakes, adaptive damping, engine management and other systems. A Ford Scorpio was therefore equipped with additional sensors to detect the use of all vehicle controls and to monitor actual vehicle performance. To accommodate the high sensor input count and the rate at which data needed to be sampled, a PC-based data-acquisition device – the AutoDAC – was developed.

The AutoDAC system offers a robust data-logging facility which is operated as a black-box facility in the vehicle, becoming enabled as soon the driver keys on. Four simultaneous video images are also captured using a four-channel VHS recording system which is interlinked with the AutoDAC to provide a common time base and file identification record. Miniature cameras are directed at the road ahead as well as at the driver from various positions within the car.

The final components of the system are computational algorithms which can receive data from the vehicle sensors and detect patterns which are consistent with those produced by a driver suffering from levels of fatigue sufficient to cause impairment or lead to impairment.

The development strategy required the collection of data from drivers completing extended journeys to validate the image processing and 'train' the neural net based algorithms. However, a major review of current and likely progress resulted in a number of significant changes to these plans at both practical (image-processing capabilities) and strategic levels (choice of fatigue measure).

The video recording system was capable of capturing usable images under relatively limited operating conditions. The wide range of ambient lighting conditions exceeded the VHS camera's ability to accommodate them. Strong sunlight also encouraged drivers to wear dark glasses, and given the extended nature of the planned sessions, it was felt unreasonable to prevent their use. In many situations, this rendered the images unsuitable for subsequent analysis although efforts were made to salvage the situation by way of 'manual' analysis. The camera position also led to the capture of images with rapidly changing backgrounds (e.g. buildings, trees, passing vehicles) as well as changes in illumination and this exacerbated the problem of locating the face and then subsequently tracking the eyes. It proved necessary to interrupt the procedure in order to confirm regularly that the system was indeed tracking the eyes and not some other component of the image and the number of processing cycles required placed upper limits on the sampling rate that could be achieved.

The review of the fatigue-detection criteria established that simple measures of eye-blink frequency alone would be insufficient to determine fatigue satisfactorily since many factors other than fatigue are known to influence blink rates. A more robust estimation would require the incorporation of eye-blink duration data as well but the available video frame rate would be inadequate to achieve this. As a consequence, a decision was made to adopt alternate psycho-physiological measures of fatigue until the image-processing performance could be developed sufficiently. Significant progress has since been made in improving the image-processing system's performance. This was largely as a result of increasing the computer's processing speed and power and with a switch to a colour system capable of detecting flesh tones in the image. The system is now capable of giving a continuous estimate of the drivers' eyelid separation – a substantially more reliable indicator of fatigue than eye-blink frequency.

The revised strategy involved using more direct laboratory-oriented methods to measure fatigue: cortical activity, heart rate, muscle tone and eye movements. It was accepted that these direct contact measures would be appropriate in the system's development even if they could not be part of a future system.

An initial programme of field trials was devised to provide the vehicle control and driver data necessary to develop the neural net. For future systems to be capable of recognizing instances of fatigued driving from a continuum of driver states it would be necessary to collect data from drivers who themselves were attempting to drive with markedly different degrees of arousal (i.e. from fully alert to very tired). This requirement for training data was a major determinant of the project's initial field work programme.

In designing the trials, two apparently opposing requirements had to be reconciled. These were the need to collect data regarding highly fatigued drivers under realistic conditions without jeopardizing the safety of the test drivers and research staff involved – familiar requirements in driver-fatigue research. As a consequence a series of three trials was designed which manipulated fatigue under controlled conditions but not on the public road system. This compromise allowed the use of highly fatigued drivers but did not place individuals at risk.

The first trial used a basic laboratory driving simulator to collect baseline psycho-physiological data from drivers experiencing levels of fatigue which could not be safely induced on public roads. The second trial used a real driving task on a private airfield and again with sleep deprived drivers. The final trial used a 150-mile motorway loop but without any sleep deprivation involved.

22.3 The field trial programme

Laboratory trials

In addition to investigating the onset of drowsiness and sleep during a simulated driving task, the first trial established a common trial procedure which was adopted with minor changes in the subsequent trials.

A total of 20 subjects were recruited from the local university community with the majority being students. Although minimal previous driving experience was required, health history and current medication were checked. The trial involved the use of the standard test vehicle and a basic driving simulator.

The simulator comprised a video display cabin in which the stationary test vehicle could be positioned and a view of the road ahead projected onto a 5 m × 3 m screen in front of the driver. The video footage was recorded during a four-hour daylight motorway drive with a 'lead' vehicle constantly positioned approximately 70 m in 'front' of the test vehicle. The driver's task in the simulator was to 'track' the lead vehicle by way of a small dot of red light projected onto the video image from a laser light source mounted on one of the simulator vehicle's front wheels. This arrangement allowed the driver to adjust the horizontal position of the projected dot using the steering wheel and the task was designed to mimic the relatively low demands imposed on a driver completing a motorway journey under light traffic conditions who needs to make only fine steering adjustments.

The subjects were invited to attend the test site in the early evening and were then given a full briefing on the procedure, task and measurements to be made. The subjects were then prepared and fitted with sensor electrodes by an electro-physiology technician recruited from a local hospital. The montage allowed the recording of four EEG channels, two periorbitally placed EOG channels and single channels for ECG and EMG. The data were recorded onto a standard eight-channel Medilog recorder with a data capture life of some 24 hours.

The subjects were then asked to complete a 30-minute familiarization session in the test vehicle. This preliminary session also provided baseline psycho-physiological data from the subjects when in an unfatigued state. The subjects were then invited to complete a sleep latency test. The procedure used was a simplified form of the conventional multiple sleep latency (Richardson *et al.*, 1978) test since the subjects were asked to attempt to fall asleep only once during the 30-minute period allowed. This test was repeated during the course of the trial and allowed a further longitudinal assessment of the subjects' underlying state of fatigue. The subjects were then allowed to return home with the request that they stay awake all night and refrain from consuming alcohol or caffeine.

The subjects returned to the test facility the following day at 9 a.m. or 1 p.m. for the full test session. The subjects were instructed to attempt to track the lead vehicle for as long as they could (up to a maximum of three hours) but to stop if they felt too distressed to continue (e.g. excessive nausea, etc.). The subjects were not allowed to listen to the radio or chew gum but they could adjust the ventilation.

The subjects were videoed and their task performance and behaviour was also monitored. If the subjects appeared to have gone to sleep (eye closure for 30 seconds and absence of task performance) they were roused. This was repeated a second time and on the third occurrence once the subjects appeared to have achieved 10 minutes of apparent sleep the session was terminated. All subjects then completed a further sleep latency test before having the electro-physiological recording sensors removed and being escorted home.

The data of only one of the 20 subjects proved to be impossible to analyse and this was due to the sensor electrodes becoming detached during the trial session. The data from the remaining subjects' Medilog recorders were viewed and analysed polygraphically by a highly experienced analyst.

The analyst completed an initial, and relatively high-level review to establish a profile for each subject which enabled the analyst to confirm that the dataset was complete, the subject's EEG response was conventional (some five per cent of the population do not show a conventional response) and that the subject had indeed lost the previous night's sleep. The overview also allowed the test session and sleep latency tests to be identified on the dataset.

The second, and more detailed, analysis of the test session data involved the examination of successive 16-second epochs from the polygraph with a single global score being given to each epoch. Although the scoring system takes account of all the electro-physiological data channels the primary defining characteristics for levels 1–3 are the muscle tone and eye movements while the EEG data are a more significant determinant of levels 4–6. Table 22.1 provides a summary of the drowsiness scale and the defining characteristics.

The drowsiness scale thus relates to, and subsumes, Rechtschaffen and Kales' (1968) standard scale for sleep with sleep stages 1 and 2–4 corresponding to the highest levels of the drowsiness index.

Comprehensive electro-physiological data were successfully recorded from 19 of the 20 participating subjects. The results indicated that subjects exhibited significant levels of fatigue behaviour within a short time in the simulator. One subject asked to withdraw after 117 minutes because they found the temperature inside the vehicle uncomfortably hot but the remaining 18 subjects completed a full session in the simulator – eight managing to survive for the full three hours. Table 22.2 shows the elapsed time from the beginning of the session to when subjects first exhibited significant levels of drowsiness or arousal as well as the total time spent in the simulator. Three key stages are included:

1. the onset of drowsiness (level 4);

Table 22.1 The drowsiness scale

Index	Arousal category
1	Active
2	Quiet waking plus
3	Quiet waking
4	Waking with intermittent alpha
5	Waking with continuous alpha
6	Waking with intermittent theta
7	Sleep stage 1
8	Sleep stage 1, 2, 3, 4 and REM

Table 22.2 Elapsed time (minutes), level of drowsiness and total session length

Subject	Level 4	Level 6	Level 7	End
1	13.60	30.13	55.20	61.33
2	0.53	5.33	11.20	64.27
3	2.40	8.53	33.07	77.33
4	3.20	28.53	113.33	124.27
5	17.87	38.40	40.53	51.73
6	20.00	21.60	22.13	182.13
7	2.40	12.80	0.00	180.53
8	7.47	10.13	28.00	181.33
9	20.00	20.00	39.47	95.20
10	12.00	13.87	24.00	111.73
11	1.33	13.07	65.60	85.07
12	38.13	43.20	74.93	116.80
13	11.20	33.87	58.13	181.07
14	16.80	17.07	0.00	179.20
15	10.13	13.87	47.20	58.93
16	2.13	6.40	73.07	181.60
17	12.00	18.40	83.20	181.07
18				
19	12.53	80.27	90.40	181.33
20	7.47	7.47	54.67	111.20
Mean	11.12	22.26	48.11	126.64
s.d.	9.15	17.88	30.85	51.50

2. the appearance of significant impairment (level 6); and

3. the onset of sleep (level 7).

Although there was considerable inter-subject variation, all 19 subjects showed a clear increase in recorded drowsiness during their test sessions. The effect was so strong that 17 of the subjects reached level 7 (stage 1 sleep) after an average of just 53.6 minutes. The onset of the effect was equally rapid with the mean time taken to exhibit level 4 effects being just 11.12 minutes and level 6 being 22.3 minutes.

The second most significant characteristic revealed in the analysed data was the episodic nature of the drowsiness response. None of the subjects showed a simple linear response with an increase in fatigue directly linked to length of time on task. The graphed output of each subject session clearly show that, although an underlying trend towards increased drowsiness may have been present, subjects typically experienced a succession of brief periods of reduced arousal followed by a return to a more alert status. The video recordings of the subjects' faces showed this process very clearly with many subjects showing, after an initial period of apparent alertness, an extended period in which they alternated between states of pronounced and more moderate drowsiness. This was evidenced in head nodding, periods of eye closure lasting several seconds and intermittent attempts to pursue the tracking task.

Figure 22.1 shows a typical plot from a subject in the simulator trials. Characteristic features are the relatively rapid onset of drowsiness and the frequent switching between alert and drowsy states. Over the brief course of the subject's participation there is a clear increase in the duration and magnitude of the episodes

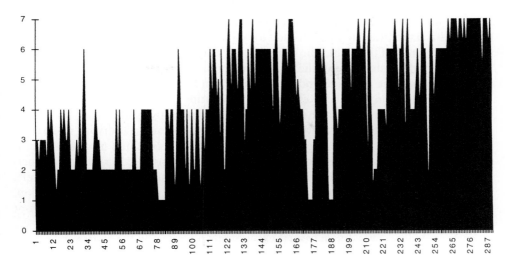

Figure 22.1 Simulator trial, subject 3: 08.48–10.05 a.m – changes in drowsiness with time (16 second epochs).

of drowsiness. The subject's session was terminated when they were judged to have fallen asleep at the wheel.

A major limitation of the first trial was the lack of an incentive for the subjects to maintain their arousal. The conditions were very conducive to sleep and with no element of 'risk' the subjects did not appear to resist sleep onset. This reduced the usefulness of the steering control data since for extended periods the drivers appeared to maintain a very reduced state of arousal with little effort to maintain task performance but without progressing to stage 1 sleep. Use of a more interactive simulator would have enabled performance-related reward schedules to have been used; see, for example, Cook, Allen and Stein (1981). The second trial was able to overcome this problem by the introduction of an element of task realism.

Closed circuit trials

The second trial was conducted on a partially used airfield and involved both day and night sessions. Two similar 'L' shaped courses of 3.5 km and 2.4 km were marked out with traffic cones and although they both had 'dead' turns at each end the width of the runway (some 50 m) allowed these turns to be made with minimal loss of speed. The drivers were instructed to attempt to maintain a speed of 30 mph and their safety was further secured through the presence of a co-driver during all the sessions who was ready to rouse the subject or take control of the steering wheel if necessary. The co-driver also made notes regarding the driver's performance and any incidents to support the interpretation of the electro-physiological data.

Some 22 subjects took part in this trial and the procedure was similar to that used in the first trial. The subjects were invited to attend the test facility in the early evening and were fitted with the recording apparatus after a full briefing. The subjects completed a 10-minute familiarization session on the track and then completed a 30-minute sleep latency test. The subjects remained on the site overnight in a private hotel and stayed awake accompanied by a supervisor. A night-time session of approximately two hours duration was completed by each subject starting at either 1 a.m. or 3 a.m. Upon completion the drivers returned to the hotel and took a second sleep latency test before staying awake for the rest of the night. The following day they completed the major test session – a four-hour drive starting at either 9 a.m. or 1 p.m. After a final sleep latency test the drivers' sensors were removed and they were driven home.

The procedure proved highly effective in generating instances of severely fatigued driving and this was evident from both subjective and objective sources of data. Examination of the daytime video recordings showed instances of subjects driving when their eyes were closed for several seconds at a time and on a substantial number of occasions the co-driver was obliged to intervene to prevent a possible loss-of-control incident. In some cases, the intervention involved the co-driver simply speaking the driver's name loudly or placing a hand on the steering wheel to make a corrective action. However, occasionally a more pronounced intervention was required, i.e. an abrupt steering wheel correction or knocking the gear lever into neutral.

Further evidence was drawn from the comments recorded on the co-drivers' record sheets. These included observations made by the co-drivers regarding the subjects' state of arousal and driving performance as well as verbal comments volunteered from the

Table 22.3 Co-driver's comment sheet from closed circuit trial (excerpt)

Time	Co-driver's Comments
08.35	Start.
08.40	Q – ? *Pretty good*.
08.55	Steering showing signs of drifting already, also told him to change into 4th gear after doing 2 laps in 3rd gear.
08.59	Subjects eyes closing, speed increasing. Q – ? *A bit tired but getting better*. He then turned heating off.
09.05	Subject asleep 10 seconds, almost missed corner.
09.08	Asleep again, speed increases.
09.11	Driver asleep, I had to wake him up, said he *felt really tired now*.
09.20	Eyes closing again, sound of (event recorder) button woke him up, subject seems to be really struggling to keep awake.
11.03	Noticed speed starting to get erratic, braking very harshly at ends of runway, no gear changing anymore.
11.07	Speed very erratic, driver asleep. Q – ? *My reactions are getting very slow but I wish to continue*. I am not too sure though (co-driver comment).
11.16	Car wandering again.
11.22	Driver falling asleep again – I think we have just about reached our limit (co-driver comment).
11.24	Spoke to driver; he did not hear me initially.
11.26	Q – ? *Not too bad, but very tired*.
11.30	Driver drifting again, eyes becoming very heavy.
11.32	Fighting to keep eyes open, seems very jumpy, speed fluctuating.
11.34	Driver asleep and speeding.
11.35	Driver asleep almost full length of runway.
11.37	Driver asleep until reaching corners, hands sliding on wheel.
11.40	Driver almost asleep passed end of runway. I feel unhappy about continuing much longer. I've decided to give it another 10 minutes.
11.44	Driver asleep full length of runway.
11.47	Driver blinking a lot.
11.48	Driver asleep, passed end of runway, shouted at him. ...
11.49	End.

subjects and their responses to regular questioning from the co-driver regarding their fitness to drive. Table 22.3 contains excerpts from a co-driver's comment sheet.

Although the driving performance varied significantly, a striking finding was the ability of some subjects to continue driving while in an extremely fatigued state. This often included driving the length of the runway with eyes closed and only regaining apparent control at the co-driver's intervention prior to the turn. In these cases although there was a tendency for the vehicle to drift slowly off course, vehicle speed was often maintained and in some cases, much to the alarm of the co-drivers, actually increased. As with the 'simulator' trials, there was marked tendency for subjects to experience episodes of almost overwhelming fatigue interspersed with periods of greater arousal (Figure 22.2). The plot is very similar to that shown in Figure 22.1 except that the duration is considerably longer and the drowsiness episodes, while just as severe, are briefer. This is presumably a response to the greater arousal provided by the task.

Road trials

The experience of the second trial clearly ruled out any possible manipulation of fatigue in the road trials but apart from allowing the subjects to sleep prior to the daytime test sessions the procedure was very similar to that employed in the second trial.

The subjects were drawn from a database of volunteer drivers maintained by HUSAT and only those with a number of year's driving experience of driving – and a significant proportion of this on motorways – were recruited.

A 120 km section of the M1 motorway between Loughborough and Luton was used in the study and the drivers had to complete one lap (240 km) at night and two laps (480 km) the following day. The drivers arrived at the test site in the early evening and were fitted with the electro-physiological recording equipment before a brief familiarization drive in the car and a first sleep latency test. The drivers then completed a lap of the test circuit starting at either 9 p.m. or 11 p.m. followed by a second sleep latency test.

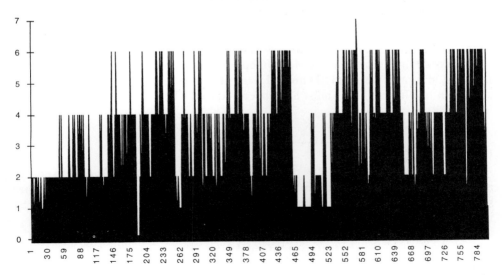

Figure 22.2   Closed circuit trial, subject 21: 08.56 a.m.–12.29 p.m. – changes in drowsiness with time (16 second epochs).

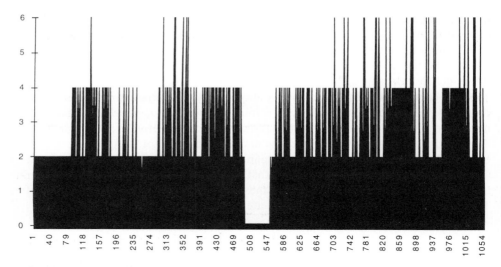

Figure 22.3 Road trial, subject 12, 13.14–17.55 p.m. – changes in drowsiness with time (16 second epochs).

The drivers then went home for a night's sleep before returning to the test site and completing a two-lap four-hour drive and sleep latency test the following day.

The four-hour session was punctuated by a vehicle re-fueling break after approximately two hours. Although no co-driver was employed the drivers were accompanied by a member of the research team who was primarily present to operate the data logging equipment and assist should a fault occur with the vehicle. However, the researcher also kept a discrete watch on the driver's apparent alertness and performance.

Most of the drivers reported an increase in the level of subjective fatigue experienced over the course of the trials. One subject's fatigue response is shown graphically in Figure 22.3. Although the level of drowsiness recorded is nothing like that found in the first two trials there are still some potentially alarming features. For a considerable portion of the journey, the subject remained in a relaxed state (level 3) but frequently switched in and out of a more drowsy state (level 4) with occasional episodes of pronounced drowsiness (level 6). Most of the subjects drove the entire route at or near the legal maximum speed (110 km) and their ability to respond quickly to a sudden change in traffic conditions must have been reduced when experiencing the highest levels of drowsiness.

Future work

At the time of writing, the off-line processing of the fatigue and vehicle data to enable net generation has still to be completed. However, once this has taken place, a further session of field trials is clearly required to assess the accuracy of the net's output. The initial validation is likely to be compared once again with a direct psycho-physiological measure of fatigue. However, once the net is capable of achieving an output, a second, and equally important, type of validation will be required. This is a direct comparison of the net's output with the drivers' self assessments of their own arousal. If drivers are unaware of subjective feelings of fatigue then this is likely to have a significant impact on their readiness to respond positively to a warning from a device. The optimal timing for warning outputs will largely depend on a comparative assessment of the drivers' objectively measured, and subjectively experienced, state of fatigue.

Acknowledgements

The author gratefully acknowledges the support of the Ford Motor Company for their support in completing the field trials.

References

Carsakadon, M.A., Littell, W.P. and Dement, W.C., 1985, Constant routine: alertness, oral body temperature and performance, *Sleep Research*, **14**, 293.

Cook, M.L., Allen, R.W. and Stein, A.C., 1981, Using rewards and penalties to obtain desired subject performance, *Proceedings of the Seventeenth Annual NASA-University Conference on Manual Control*, Pasadena, CA: Jet Propulsion Laboratory.

Hamelin, P., 1987, Lorry drivers' time habits in work and their involvement in traffic accidents, *Ergonomics*, **30**, 1323–33.

Harris, W., 1977, Fatigue, circadian rhythm and truck accidents, in Mackie, R.R. (Ed.) *Vigilance*, New York: Plenum Press.

O'Hanlon, J.F., 1978, What is the extent of the driving fatigue problem? *Driver Fatigue in road accidents*, Commission of the European Communities, EUR 6065 EN.

Rechtschaffen, A. and Kales, A., 1968, A manual of standardised terminology techniques and scoring system for sleep stages of human subjects, Washington, DC: US Government Printing Office.

Richardson, G.S., Carsakadon, M.A., Flagg, W., van den Hoed, J., Dement, W.C. and Mitler, M.M., 1978, Excessive daytime sleepiness in man: Multiple sleep latency test in narcoleptic and control subjects, *EEG Clinical Neurophysiology*, **45**, 621–7.

SECTION FIVE:
Theoretical considerations in research into driving

23 Dysfunctional driving behaviours: a cognitive approach to road safety research

John M. Reid

23.1 Aim of the research

Throughout the world in countries where there is a substantial volume of vehicular traffic, the incidence of road collisions, and the resulting deaths, injuries and property damage, is regarded as a significant social problem. Although there are studies that identify driving maneouvres that are more or less risky (Charlesworth-McIntyre, 1988) and many studies that describe the errors young drivers make, (Catchpole *et al.*, 1992; Milech *et al.*, 1989) there are very few, if any, studies that address the question: 'Why did the driver do what he or she did that led to the collision?' From a cognitive psychological perspective, 'Why?' implies 'What motivated the behaviour?', and that question is equivalent to asking: 'What was the emotion that motivated the behaviour?' Consequently, the aim of this research is to investigate the motivations for dysfunctional driving behaviours (DDBs). (Collisions where alcohol or other substances are a factor, and intentional collisions where suicide or homicide are motivations are not covered by this research. The studies reported here represent the preliminary stage of the project; further studies of the effects of emotion on attention and the structure and contents of driving schemata are in progress.)

This chapter reports the first stage of the research, during which a taxonomy of DDBs was developed and members of the Victoria Police who had been involved in collisions were interviewed to validate the taxonomy. It is expected that products of the research will include a cognitive model of driving behaviour and a prescriptive model of safe driving. These will provide empirically derived, theoretic foundations for a number of practical applications in the fields of road safety policy, planning and implementation, including driver training. It is also expected the research will provide a model for designing a diagnostic test of people's attitudes towards driving behaviours.

23.2 A Taxonomy of DDBs

Categorizing the affect-motivated behaviours that increase the probability that a person will be involved in a collision was the starting point for the research. A list of driving actions that appeared to be dysfunctional was produced by observing drivers' road behaviours. As a step towards validating the list, expert, experienced drivers were consulted to obtain their opinions about the comprehensiveness of the list and the appropriateness of the categories. Further face validity was obtained by interviewing Victoria Police drivers who had been involved in collisions and asking them about the circumstances of their collisions, including their affective and cognitive states immediately prior to the incidents.

The information produced by these studies resulted in the proposal that there are six categories of affect-motivated behaviours involved in collisions:

- aggressive-competitive behaviour,
- risk-taking behaviour,
- drowsiness,
- misattention,
- faulty schemata, and
- carelessness,

and these are sufficient to explain the behaviours that resulted in all the collisions investigated.

Aggressive-competitive and risk-taking behaviours

Aggressive-competitive behaviour (A-CB) and risk-taking behaviour (R-TB) are conscious, strongly affect-motivated behaviours. They may be exhibited as relatively transient responses to emotions activated by neural, sensorimotor, affective, or cognitive processes (Izard, 1993), or they may be manifestations of a driver's core constructs – hence, they may be represented as characteristic driving styles.

A-CB is always directed against another person and it is motivated by negative emotions, e.g. frustration, anger and/or contempt. R-TB is goal-directed and it may be motivated by positive (e.g. exhilaration, joy) or by negative (e.g. frustration, distress, anger) affect.

Drowsiness

Drowsiness is a state of diminishing arousal in which the organism is approaching the threshold of sleep. The relationship between diminished arousal (as measured, for example, by decreased blood/urine levels of catecholamines) and sustained attention (vigilance) and the performance of complex tasks is the subject of a considerable body of research literature (Parasuraman, 1984). It is beyond the scope of this chapter to discuss the psychopharmacological relationship between arousal and vigilance, although the effects of adrenergic chemicals on arousal (Davies and Parasuraman, 1981; O'Hanlon and Beatty, 1976; Frankenhäuser, 1975) seem to be particularly relevant to police and other emergency drivers.

Discussing the effects of decreased arousal on the performance of complex tasks, Parasuraman (1984) points out:

> Tasks requiring sustained attention, being prolonged and often monotonous, may sometimes induce drowsiness or feelings of boredom in the subject ... Vigilance decrement is not a unitary process ... Lowered arousal is clearly a concomitant of the decrement (in the performance of such tasks), and it probably plays a contributory role in tasks performed under extremes of arousal, such as after a period without sleep or after ingestion of a stimulant.

and he suggests that other CNS functions are also involved (ibid: 64).

Fatigue is one of the causes of reduced arousal, but circadian rhythms (Harris, 1977; Horne, 1992) and shift-working are probably also factors involved in performance decrement. Depression has been reported as being a factor in poor performance on vigilance tasks (Byrne, 1976). Narcolepsy may also cause some drivers to fall asleep at the wheel. Whatever the cause of drowsiness, it is a state in which emotional arousal is ebbing away to a dysfunctionally low level, the perceptual monitoring systems are inhibited and the driver ceases to respond appropriately to environmental stimuli.

Misattention

The fact a driver may not be attending to the driving task does not necessarily represent dysfunctional behaviour. An experienced driver can hold a conversation with a passenger, listen to the car radio, or think about the meeting he or she will be attending later that day, and automatically perform the driving actions efficiently and safely. If something in the environment requires attention (usually something that represents a potential hazard), the driver's attention will switch from the distractor to the stimulus. The driver will assess the stimulus and implement an appropriate response.

However, if the driver is concentrating too strongly on a distractor, his or her attention may not be captured by a potential-hazard stimulus. As implied by Parasuraman (1984), attention and emotional arousal are concomitants; the greater the intensity of emotion, the more strongly attention is focused on the stimulus that is evoking the emotion. Being emotionally aroused towards a distractor to such a degree that the driver fails to orient to a potential hazard is described as misattention. This includes the 'human error' sometimes referred to as 'looked but failed to see' (Sabey and Staughton, 1975) which is simply a manifestation of the same phenomenon.

Obviously, if a driver is not scanning the environment, whether automatically or consciously, hazards will not be seen and the driver will not implement appropriate hazard-avoidance actions. This behaviour is also categorized as misattention.

Faulty schemata and carelessness

Every activity an organism engages in is represented by a schema, and schemata are learnt by engaging in the activity. When people are taught to drive correctly, the schemata they form represent functional driving behaviours. A driver may form a faulty driving schema either because he or she has been taught dysfunctional driving behaviours or they may have learnt such behaviours heuristically. For example, turning right at an intersection is a manoeuvre that is often performed dysfunctionally. Instead of following a trajectory that keeps the vehicle in the correct lane, drivers often 'cut the corner'. Assuming they were taught to perform the manoeuvre correctly, why do so many drivers often perform it dysfunctionally? Probably because performing the manoeuvre correctly requires that it be executed relatively slowly and with some effort. In other words, it is easier to cut the corner than to go round it. The careless behaviour in this case is motivated by laziness.

In another case, perhaps the driver is turning right at an intersection across a stream of on-coming traffic. Instead of waiting for a sufficiently long gap in the stream to allow the turn to be performed correctly, the driver cuts the corner in order to hurry the manoeuvre. In this case, Carver and Scheier's (1990) control-process model would suggest the motivation for the dysfunctional behaviour could be negative affect resulting from the driver's construal that his or her progress towards a goal is 'at a rate lower than (a desired) standard' (Carver and Scheier, 1990).

In both the above examples, it is implied that the dysfunctional manoeuvre is performed consciously – or, at least, that on some occasions the driver performs the manoeuvre correctly and it is only when motivated by negative emotion (laziness or frustration) that the dysfunctional behaviour occurs. Such behaviours are assigned to the carelessness category of DDBs. Carelessness, as defined for this context, represents the situation where a driver behaves in a way that is discrepant with their (functional) driving schemata.

However, after a manoeuvre has been performed carelessly on a number of occasions without incurring any penalty (i.e. without resulting in an actual or near collision or the issuing of a penalty infringement notice), and possibly the driver is rewarded by a reduction in the goal-attainment discrepancy, it may become habitual. In this case, the driver's schema changes to incorporate the dysfunctional behaviour and he or she unconsciously always performs the manoeuvre incorrectly. This DDB category is described as faulty schemata.

When a well-trained driver is attending to a distractor, but is oriented so that the road ahead is within the visual field, he or she also scans the environment (and this should include the rear-vision mirror) automatically. When a potentially hazardous stimulus (PHS) that is represented as a slot in the schema (Mandler, 1984; Schank and Abelson, 1977 – see example, below) for the particular driving activity captures a driver's attention (i.e. when he or she 'orients' to the PHS), a decision-making process is initiated. The process involves assessing the PHS, including making a subjective probability judgement of the likelihood that an actual hazard will materialize, and preparatory behavioural set, rather than an immediate automatic motor response. On the other hand, if the hazard is an actual threat in the immediate vicinity of the driver, the startle reflex may be triggered and an automatic (non-cognitive) reaction will be effected.

The schema that is salient is the analogue of whatever the organism is attending to (i.e., is conscious of). If that schema is a distractor, a schema of the driving action that is being performed automatically monitors the driving process and feedback from the environment. The driver's attention is oriented to the driving activity if a sensory input is discrepant from the subconscious schema, or if the input represents a slot in the schema that identifies a stimulus that must be attended to – that is, a potential hazard. (It may be that, *a priori*, the probability of a hazard eventuating has to reach a threshold before the stimulus will result in attentional switch.)

A hazard, such as a child running out onto the road from in front of a vehicle, may be cued by a PHS, such as a school bus parked in front of a school, or a school-crossing. As suggested above, the PHS represents a slot in the driver's schema for driving-along-a-street-in-which-a-school-is-located; if the driver's schema does not contain such a slot – in other words, if the driver is not aware of the danger that a child may emerge from in front of a school bus – the hazard will not be cued and the latency of the driver's orienting response may be increased (e.g. Yantis and Jonides, 1990; see also D.A. Norman's comment in Posner, 1984: 50). This absence of a hazard-representative slot is another case of a faulty schema.

Note that if a PHS is not part of a driver's schema, the abrupt onset of an actual hazard may trigger the startle reflex, rather than an orienting response, and the automatic collision-avoidance reflex behaviour may well be sub-optimal in particular circumstances.

23.3 Interviewing Victoria police drivers

The process of interviewing Victoria police drivers who have been involved in collisions is continuing and the preliminary results of 64 interviews are included, here. It should be noted that in most cases, when other drivers were involved, these interviews represent only one side of each story, and often the police member appeared to be less responsible for the collision than the other party. Nevertheless, in almost all cases, the interviewees conceded they could have acted differently and thereby avoided the collision. This issue is discussed further, below.

Method

Subjects

The subjects were 60 male and four female members of the Victoria police who had been involved in collisions that resulted in their approved driver's certificates being suspended. Initially, interviews were conducted with members whose collisions occurred up to two years before the interview; latterly, however, members have been interviewed immediately prior to their attending the defensive driving course at the Victoria Police Motor Driving School, usually within six months of the collision. The ages of the members interviewed range from 23 to 55 years.

Procedure

The interviews were in two parts. The first part was based on the critical incident technique (Flanagan, 1954). Interviewees were asked to describe the circumstances of their collision, including how they were feeling and what they were thinking about immediately prior to the collision. The interviewer probed to elicit as much detail as possible, including challenging what appeared to be inconsistencies or improbabilities in a member's account of the incident. Having satisfied himself that the member had given a full and relatively uncoloured description of the circumstances of the collision, the interviewer asked the member to say what he or she could have done to avoid the collision. Each interview lasted for approximately one hour. In all but one case, the interviewees agreed to the interview being recorded.

The second part of the interview consisted of completing a rank order repertory grid (Fransella and Bannister, 1977), in which both the elements and the constructs, including the contrast poles, were supplied. The elements (roles) comprised three 'approved-of' and three 'not-approved-of' drivers:

A: Someone I feel comfortable being driven by

B: Someone I do not feel comfortable being driven by

C: Someone I admire as a driver

D: Someone I do not admire as a driver

E: Someone I regard as a very competent driver

F: Someone I do not regard as a competent driver

Interviewees provided the names of six people they knew who fitted the roles and these names were written on cards. The elements also included three manifestations of 'self':

G: Me as a driver as I am now

H: Me as a driver as ideally I would like to be

I: Me as a driver as I was when I was driving on a P-plate

The constructs (driving styles) were:

- a cautious driver who seldom takes risks
- a driver who is always trying to get ahead of other drivers
- a vigilant driver whose attention is not easily diverted
- a careless driver – for example, the sort of person who cuts corners
- an alert driver who would never drive when drowsy, and
- a driver who doesn't plan ahead and reacts to situations as they arise.

The members ranked the cards on each of the constructs to produce 9×6 matrices.

Results

Interviews

The interview protocols were analysed to determine whether the behaviours that appeared to be precursors to the collisions could be categorized according to the taxonomy of DDBs. Because the interviews were necessarily relatively brief and because it was not feasible to visit the scene of the collision, or speak to others involved, in some cases deciding which category of DDB to assign an incident to (e.g. *misattention* or *faulty schema*) involved a probability judgement. However, all the incidents could be comfortably accommodated within the taxonomy of DDBs.

In three cases, when the police vehicles were rammed from behind or deliberately rammed by an offender, it appeared there was nothing the police driver could have done to avoid the collision, hence, their behaviours were not dysfunctional. Of course, this does not mean these collisions were due to causes that do not fit the taxonomy of DDBs, but since the other drivers could not be interviewed, it was not possible to categorize the incidents.

In two other cases, the interviewees maintained the collisions were caused by the brakes of their vehicles locking, although both admitted there was no evidence of a malfunction when the vehicle was later inspected. It may have appeared from the circumstances of the collision there were other more likely causes, but because the interviewees steadfastly maintained there was nothing they could do to prevent the collision from occurring – one of them said, 'I did everything right, but the brakes locked up and there was nothing I could do about it!' – they could not be categorized as DDBs, although the fact the interviewees were in situations where it was necessary to brake hard may be regarded as significant.

Several other interviewees tentatively advanced the 'brakes locked hypothesis' as a contributory factor to account for their collisions – 'I can only think the brakes must have locked' – but in all cases they were able to say how they could have acted differently

Table 23.1 Number of collisions in each category of DDB

Category of DDB	Number of collisions (%)	
Faulty schema	17	(27)
Misattention	16	(26)
Carelessness	12	(19)
Risk-taking behaviour	8	(13)
Aggressive-competitive behaviour	5	(8)
Drowsiness	1	(2)
No fault by police driver	3	(5)
Total	62	(100)

and thereby have avoided the collision; consequently the incidents were assigned to what appeared to be the appropriate categories of DDB.

The analyses of the interview protocols indicate that having faulty schemata or misattention were the principal factors in the majority of cases, followed by carelessness. Table 23.1 shows the numbers of collisions assigned to each category of DDB.

Across the categories, three common factors were present in a number of collisions: operations involving lights-and-sirens, or silent pursuits culminated in collisions in 18 cases (i.e. 29 per cent of 62 collisions); failure to have the observer dismount when reversing resulted in seven collisions; and what we refer to as the 'empty lane fallacy' (see 23.4 Discussion) resulted in nine collisions.

Repertory grids

The repertory grids of 30 interviewees whose collisions occurred within the last six months were analysed in two ways:

1. the mean of the ranks for each role on each construct was calculated and plotted as graphs (Figures 23.1–23.4); and

2. the data were factor analysed by the principal components method (Tables 23.2 and 23.3; Figures 23.5 and 23.6).

Figure 23.1 shows the mean ranks for the three approved-of roles:

- someone I feel comfortable being driven by (A),
- someone I admire as a driver (C), and
- someone I regard as a very competent driver (E).

Apart from C, who is vigilant and plans ahead, and A, who tends to be cautious, the approved-of roles are middle-of-the-road on all constructs.

Figure 23.2 shows the mean ranks for the three not-approved roles:

- someone I do not feel comfortable being driven by (B),
- someone I do not admire as a driver (D), and
- someone I do not regard as a competent driver (F).

Two aspects of this graph are particularly noteworthy; the three not-approved roles are

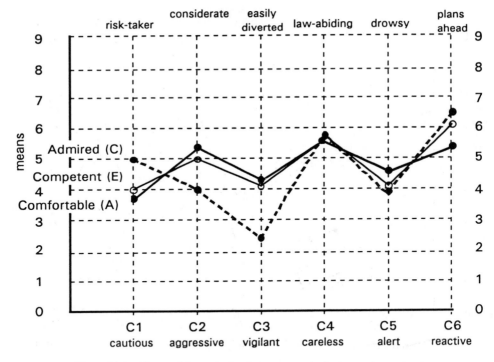

Figure 23.1 Means of the ranks for three approved-of roles on six constructs.

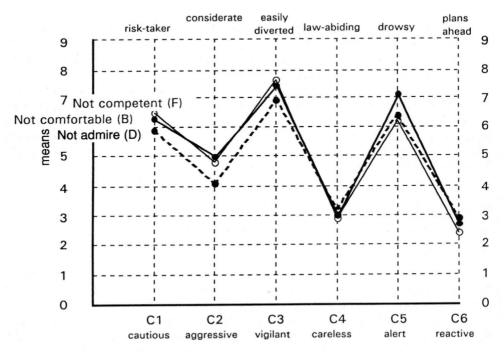

Figure 23.2 Means of the ranks for three not-approved roles on six constructs.

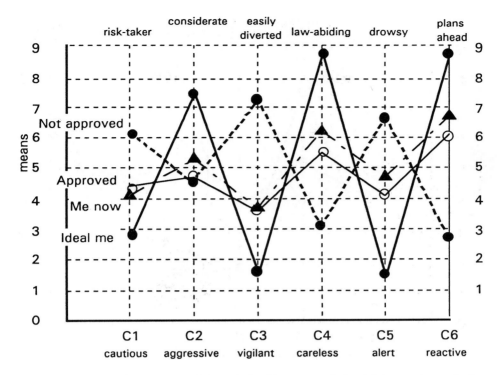

Figure 23.3 Averages of the means of the ranks for three approved-of and three not-approved roles compared with the means of the ranks for three me roles on six constructs.

seen as more similar than the three approved-of roles, and their attributes are more extreme on all constructs, except on the aggressive–considerate dimension. The not-approved roles are easily diverted and do not plan ahead; they are careless, inclined to drive when drowsy, and tend to be risk-takers.

Figure 23.3 compares the average mean ranks for approved-of and not-approved-of, with 'me now', and 'ideal me' roles. 'Me now' is very similar to the average approved-of, and both are the antithesis of the average not-approved-of. However, 'me now' and approved-of are much less well-defined than the 'ideal me', who is extremely law-abiding and anticipates extremely well. However, 'ideal me' is not as considerate of other road-users, nor as cautious, vigilant and alert as one might expect.

Figure 23.4 shows the relationships between the three 'me' roles:

- me as a P-plate driver is similar to not-approved-of,
- me now is intermediate between not-approved-of and
- ideal me.

The factor analyses of the data are shown in Tables 23.2 and 23.3, and as scattergrams, Figures 23.5 and 23.6. Table 23.2 gives the factors and loadings from the analysis of the roles (elements) matrix. Factor 1 accounts for 93 per cent of the variance; only someone I feel comfortable being driven by (A) and someone I admire as a driver (C) make any substantial contribution to Factor 2, and these roles are inversely related. From the graphs in Figures 23.1–23.4, it can be seen that the greatest divergence between these

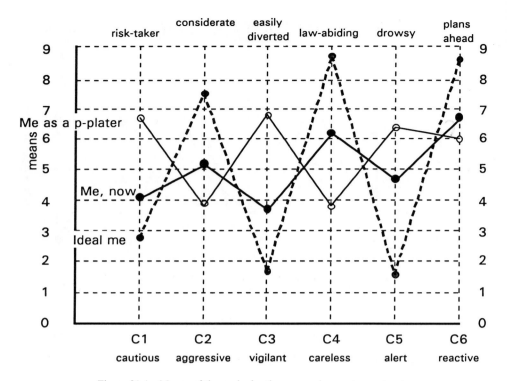

Figure 23.4 Means of the ranks for three me roles on six constructs.

Table 23.2 Factors and loadings of elements (roles) and proportionate variance contributions of the two factors

Elements	Factor 1	Factor 2
A: Someone I feel comfortable being driven by	0.8704	−0.4598
B: Someone I do not feel comfortable being driven by	−0.9858	−0.1201
C: Someone I admire as a driver	0.816	0.5718
D: Someone I do not admire as a driver	−0.9919	−0.0066
E: Someone I regard as a very competent driver	0.9792	−0.0182
F: Someone I do not regard as a competent driver	−0.9904	−0.0993
G: Me as a driver as I am now	0.9774	0.0303
H: Me as a driver as ideally I would like to be	0.9708	−0.1055
I: Me as a driver as I was when I was driving on a p-plate	−0.9665	0.2058
Proportionate variance contributions:		
Orthogonal Direct	0.9296	0.0704
Oblique Direct	0.6051	0.4237
Joint	−0.0132	−0.0156
Total	0.592	0.408

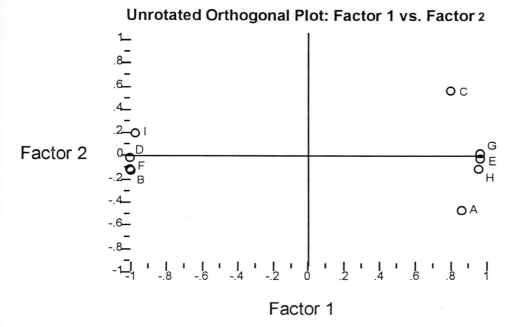

Figure 23.5 Scattergram of elements.

two roles is on the vigilance dimension, although they diverge somewhat on the aggressive and cautious dimensions. Scattergram, Figure 23.5, shows the relationships between the elements.

The factor analysis of the constructs matrix is shown in Table 23.3. Again, Factor 1 accounts for over 90 per cent of the variance. Aggressive is the only construct that shows a relatively weak loading on Factor 2. Scattergram, Figure 23.6, illustrates the relationships between the constructs.

Table 23.3 Factors and loadings of constructs and proportionate variance contributions

	Factor 1	Factor 2
Cautious	0.9526	−0.0584
Aggressive	−0.7651	0.6394
Vigilant	0.9458	0.3075
Careless	−0.9926	−0.0228
Alert	0.9753	0.061
Reactive	−0.9772	−0.1758
Proportionate variance contributions		
Orthogonal Direct	0.9069	0.0931
Oblique Direct	0.795	0.3047
Joint	−0.0003	−0.0994
Total	0.7947	0.2053

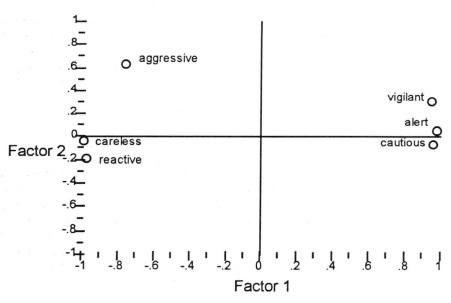

Figure 23.6 Scattergram of constructs.

23.4 Discussion

Interviews

The impact of emotion on attention stands out as the most significant factor in collisions in which police drivers were involved. In all cases, where misattention appeared to be the most likely cause of the collision and the driver was performing urgent duty, interviewees stated the 'adrenalin was pumping'. Particularly in relation to pursuit and surveillance operations, several interviewees remarked that: 'When the adrenalin is pumping, you get sort of tunnel vision.'

The effect of attending to an emotionally arousing endogenous distractor is illustrated by four cases categorized as misattention where the police members were engaged on surveillance duties. The circumstances of all four cases were so similar that one will serve to explain the effect. The member was following a suspected drug dealer, who was driving along a road. The suspect stopped and the member 'drove over' so as not to alert the target. The member then had to get himself into a position from where he could resume surveillance. He knew there was a convenient hotel carpark where he could position himself to watch the suspect and he was rehearsing the schema, the-best-way-to-get-back-the-carpark, as he began to execute a U-turn. He claimed, and there is no reason to doubt him, that he looked in his side mirror and over his right shoulder but did not see a vehicle approaching from behind until it struck him.

Some older police members found it difficult to accept they could only attend to one thing at a time. They claimed they were trained to attend to 'a number of things at once'. They were reluctant to accept that although it is possible to do several things at the same time, and attention can switch relatively quickly (say, about 300–500 ms), some cognitive scientists would say it is not possible to attend to a message coming over the police radio, think about the implications of the message, and be aware that a vehicle is about to emerge from a side road, all at the same time. It is also interesting to note that processing language has attentional priority over non-language images. One of the implications of this fact is that attending to police radio messages is likely to strongly inhibit the orienting response to a visual PHS. Even when a driver is at a comfortable operating level of arousal (as associated with interest, for example) the emerging car may not capture his or her attention. When circumstances are such that the driver is highly aroused, as is the case when a police driver is operating with lights and sirens, it is unlikely he or she will orient to the potential hazard.

In nine cases, there appeared to be a common factor in the collisions, which we refer to as the 'empty lane fallacy', representing a subset of the faulty schemata category. The drivers entered intersections where the cross traffic was stationary, but where there was an apparently empty lane down which they could not see. Instead of edging forward until they were sure the lane was empty, the drivers made the false assumption that what they could not see did not exist, and the vehicle that was travelling unseen in the lane collided with the police vehicle. In five of these cases, the driver was operating with lights and sirens and was in a highly aroused state that probably contributed to the dysfunctional behaviour – the so-called 'tunnel vision' effect. In all these five cases, the drivers were facing red lights. As they approached the intersections they slowed to a walking pace, or stopped, to check that all the cross traffic they could see had stopped to allow them to cross. They then accelerated into the intersection. All nine interviewees agreed their crossing-an-intersection schema now incorporates a slot labelled 'the potential hazard of an apparently empty lane'.

Repertory grids
One of the problematic aspects of rank order grids is that forcing subjects to rank the elements tends to mask the true distances (or similarities) between the role concepts. In an attempt to overcome this shortcoming, the means of the ranks were used and the resulting scales treated as interval level data for the factor analysis. Of course, the

procrustean effect was built into the original rank orders, so using the means (instead of the rank order of the means) may be a somewhat artificial device.

The results reported above show the Victoria police drivers interviewed have clear concepts of what represent functional and dysfunctional driving behaviours (as evidenced by the 'ideal' driver's profile) but they recognize that they themselves do not always behave, as drivers, as well as they might. What is more, drivers they approve of are very similar to themselves, neither particularly good nor particularly bad drivers – only the person they admire as a driver diverges from the middle of the road on two constructs; this person tends to be vigilant and not easily diverted from the driving task.

How the interviewees construed dysfunctional behaviours is better defined than their construal of functional behaviours. Drivers they do not approve of are easily diverted, inclined to drive when drowsy and are reactive (i.e. do not display good 'forward observation'). Aggressive-competitive behaviour, as opposed to being considerate of other road-users, does not seem to be particularly relevant to their concepts of good and bad driving behaviours. This observation is perhaps borne out by the factor analysis, which shows that all the constructs load on Factor 1, but that the construct, aggressive – considerate also loads on Factor 2.

23.5 Conclusions

The clearest general conclusion so far to emerge from this research is that high levels of emotional arousal and safe driving are inimical. Analysis of the interview protocols suggests the following generalizations:

1. when a driver's attention is focused on the driving task, high arousal commonly leads to risk-taking behaviour, or sometimes to aggressive-competitive behaviour; and
2. when a driver's attention is focused on a distractor, high arousal leads to misattention.

The message incorporated into a number of road safety public awareness campaigns, 'Speed kills!' should, perhaps be changed to 'Emotion kills!' In this context, it might be noted that 'speed' is usually simply the expression of risk-taking behaviour, and perhaps there is a correlation between level of emotional arousal and speed.

It is also seems clear that an understanding of the cognitive processes, attention and schemata formation, is necessary to devising strategies that might substantially reduce road collisions. Although emotion, attention (including automaticity and control), and schema theory are often regarded as separate areas of cognitive psychological research, at least in the context of driving, they need to be brought to a common focus on the problem. The next stage of the present research will show how emotional arousal evoked by a distractor affects the latency of the orienting response to a hazard that is cued, or not cued, depending on whether the subject's driving schemata incorporate the relevant potential hazard slots.

The research has so far investigated the circumstances of police collisions, but there is no reason to believe the general conclusions are not applicable, *mutatis mutandis*, to ordinary drivers. Emergency drivers are subject to particular emotionally arousing stimuli (e.g. lights and sirens) that other drivers do not normally experience; nevertheless, a driver using a telephone, whether 'hands-free' or not, may be as distracted as a police driver responding to a radio message, and a highly animated discussion, whether acrimonious or enjoyable, between a driver and a passenger may inhibit the driver's response to a potential hazard.

Acknowledgements

We wish to thank the Victoria Police for their assistance in carrying out this research and, particularly, the individual members who agreed to being interviewed for this study.

References

Byrne, D.G., 1976, Vigilance and arousal in depressive states, *British Journal of Social and Clinical Psychology*, **15**, 267–74.
Carver, C.S. and Scheier, M.F., 1990, Origins and functions of positive and negative affect: a control-process view, *Psychological Review*, **97.1**, 19–35.
Catchpole, J., Cairney, P. and Macdonald, W., 1992, Young drivers – what are they doing wrong? Paper presented to the Road Safety Researchers Conference, University House, Canberra, Australian Road Research Board, WD RS92/024.
Charlesworth-McIntyre, K, 1988, Unsafe driving actions, in *Driver behaviour: a joint Australian Psychological Society and Victoria Police initiative*, Proceedings of the Colloquium held at the National Sciences Centre, Parkville, Victoria, 6 December 1988, Melbourne: Victoria Police.
Davies, D.R. and Parasuraman, R., 1982, *The Psychology of Vigilance*, London: Academic Press.
Flanagan, J.C., 1954, The critical incident technique, *Psychological Bulletin*, **51.4**, 327–58.
Frankenhäuser, M., 1975, Experimental approaches to the study of catecholamines and emotion. in Levi, L. (Ed.), *Emotions their parameters and measurement*, New York: Raven.
Fransella, F. and Bannister, D., 1977, *A Manual for Repertory Grid Technique*, London: Academic Press.
Harris, W., 1977. Fatigue, circadian rhythm, and truck accidents, in Mackie, R.R. (Ed.), *Vigilance: theory, operational performance and physiological correlates*, New York: Plenum Press.
Horne, J., 1992, Stay awake, stay alive, *New Scientist*, **133**, 20–4.
Izard, C.E., 1993, Four systems for emotion activation: cognitive and noncognitive processes, *Psychological Review*, **100.1**, 68–90.
Mandler, J.M., 1984, *Stories, Scripts, and Scenes: Aspects of Schema Theory*, Hillsdale, NJ: Lawrence Erlbaum.
Milech, D., Glencross, D. and Hartley, L., 1989, Skills acquisition by young drivers: Perceiving, interpreting and responding to the driving environment, Report MR4, Canberra, ACT: Federal Office of Road Safety.
O'Hanlon J.F. and Beatty, J., 1976, Catecholamine correlates of radar monitoring, *Biological Psychology*, **4**, 293–304.
Parasuraman, R., 1984, The psychobiology of sustained attention, in Warm, J.S. (Ed.) *Sustained Attention in Human Performance*, Chichester: John Wiley & Sons.
Posner, M., 1984, Current research in the study of selective attention, in Donchin, E. (Ed.) *Cognitive Psychology: Event-Related Potentials and the Study of Cognition. The Carmel Conferences*, Vol. 1, Hillsdale, NJ: Lawrence Erlbaum.
Sabey, B. and Staughton, G., 1975, Interacting roles of road environment, vehicle and road user, in *Proceedings of the Fifth Conference*, 1–5 September 1975, London: International Association for Accident and Traffic Medicine.
Schank, R.C. and Abelson, R.P., 1977, *Scripts, Plans, Goals and Understanding*, Hillsdale, NJ: Erlbaum Associates.
Yantis, S. and Jonides, J., 1990, Abrupt visual onsets and selective attention: Voluntary versus automatic allocation, *Journal of Experimental Psychology: Human Perception and Performance*, **16.1**, 121–34.

24 Driving simulation in Australia: opportunities for research and application

Thomas J. Triggs, Alan E. Drummond and John Stanway

24.1 Introduction

While aviation simulation technology still leads the way, there has been limited use of high-end (i.e. multi-million dollar) ground vehicle simulators for a period approaching two decades. However, there has been a growing momentum in ground vehicle simulation over the past two to three years which has seen very rapid advancement of technological capabilities, and a consequent reduction in unit prices, to the stage where access to simulation technology has potentially become a general tool for road safety research and training.

To some extent, this development necessitates a 'paradigm shift' for Australian road safety, at both the research and policy levels, for it ushers in a 'new', but complementary, way of approaching the research and development of road crash countermeasures. This chapter provides some contextual information on the 'new road safety' and outlines some potential applications of simulation technology.

Background issues

There can be no doubt that over the last 20 years or so very significant road safety progress has been achieved and, while not solely responsible, the Australian road safety programme has been both effective and cost-effective. For all of this time, road safety problems have been identified largely through a crash-driven process. It can be said that mass crash data have ruled the road safety roost. There have been three major implications of the 'crash model' for road safety research: a descriptive approach; restricted road safety topics; and effectiveness vs efficiency.

A descriptive approach

Road safety, as an applied discipline, is still largely atheoretical and much road safety endeavour is limited to the descriptive domain. From a public health perspective and with very large crash frequencies (relative to today) to address, this has not been perceived as a problem in the past. However, as continued road safety progress becomes harder to achieve, the limitations of the traditional road safety model will become more apparent. This is nowhere more clearly illustrated than in the area of young driver safety. Traditionally, the apparent importance of young driver safety has been in inverse proportion to the effort to understand the problem(s). Despite the fact that there has been virtually no success in specifically improving young driver safety, there is still a widespread and firmly held view that continuing to describe the young driver problem is sufficient to develop effective and efficient crash countermeasures. Consequently, the young driver literature is a mix of factors (age, experience, personality, risk taking behaviour, etc.) and 'solutions' (training, licensing, education, exposure reduction, etc.) often advanced solely on the basis that young drivers are highly overrepresented in road crashes.

How a road safety problem is viewed determines what the road safety system attempts to do about it. Consequently, if the problem is thought to be simple or straightforward, simple or straightforward solutions will often be proposed. It is not surprising, therefore, that the most common response to the question, 'What can we do about the young driver problem?', takes the form, 'More driver training!' (regardless of the decades of failure in this area). The fact that the young driver problem has shown itself to be one of the most intractable road safety problems is a reflection of the complexity of the problem relative to other road safety issues.

Restricted road safety topics

The heavy reliance on mass crash data has also restricted the range of questions which are considered to be valid road safety questions. Historically, a valid road safety question is one which corresponds directly to mass crash data, and therefore can have crash frequencies assigned to it, and also deals with a specific issue.

Thus, there is a wealth of information on crash trends, where crashes occur, how serious the injuries are, when the crashes happen, what type(s) of road user(s) are involved, what the characteristics of these road users are, what types of vehicle are involved, etc. When a number of these factors are linked, the analysis outcome often assumes much greater importance in the research and development process than it perhaps deserves.

While there has been some excellent experimental work, such a focus on (driving) performance breakdown, that is, crashes, has meant that the formal road safety system is unable to make substantive comment on driving performance itself, perhaps the core process in road safety and one which should be central to a relevant and comprehensive road safety programme.

Effectiveness vs efficiency

Traditionally, the three 'E's in road safety have stood for education, enforcement and engineering. In terms of countermeasure assessment and evaluation, road safety has concentrated on just one further 'E', that is, effectiveness. However, there is at least one further 'E' which warrants consideration, efficiency; another 'E', equity, is a more difficult concept to incorporate into crash countermeasure design.

The usual form of countermeasure evaluation has focused on effectiveness, i.e. following an intervention, has there been a reduction in the incidence and/or severity of crashes and, most importantly, can this reduction be attributed to the intervention rather than extraneous factors? Thus, the conclusion from a countermeasure effectiveness evaluation may be, 'This intervention reduced Type X crashes by 10 per cent.'

It may be, however, that the countermeasure was actually 50 per cent effective, but because it only had a 20 per cent overlap with the targeted problem, a crash reduction of only 10 per cent was produced. Such a countermeasure could perhaps be considered to be effective but not particularly efficient. It should be noted that assessment of the level of efficiency depends on the actual homogeneity of the problem, both in terms of problem content and the extent to which this problem(s) applies to subgroups of the

Driving simulation in Australia: opportunities for research and application 251

targeted road user group. The extent to which any particular problem applies to the whole target group could be considered to measure the equity of the intervention. These factors emphasize the desirability of a greater understanding of road safety problems.

The hypothesized relationship between problems and countermeasures is shown diagrammatically in Figure 24.1.

In summary, the potential advent of simulation technology to assist in the countermeasure development process has wide-ranging implications for road safety as a discipline. While the traditional road safety model will (and should) continue to be applied, ground vehicle simulation technology makes available for perhaps the first time the means to investigate validly, reliably and efficiently a range of road safety issues and produce a new generation of road safety initiatives.

24.2 Research applications

Introduction
Proportional to their level of fidelity and sophistication, driving simulators have the

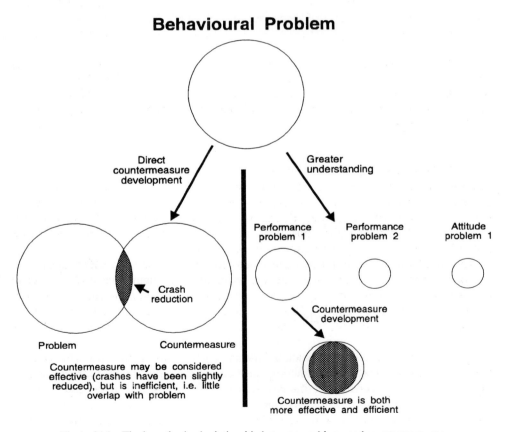

Figure 24.1 The hypothesized relationship between problems and countermeasures.

capacity to address validly any road safety issue involving driver performance. Simulation will have the potential to provide understanding of and, more importantly, solutions to some questions such as:

- What should young drivers be trained to do in order to reduce their very high risk of crash involvement?
- What effect do various forms of driver impairment have on driving?
- What is the most effective way to reduce travel speeds in residential streets?
- What types of information affect driver decision-making at intersections?
- What performance problems are reflected in the driving of elderly drivers?
- What is the best way to alert drivers to the presence of a rural rail crossing?

Rather than (real-world) trialling and evaluation, driving simulators provide the capacity to investigate a wide range of road safety areas effectively, efficiently and with a much higher degree of experimental control (for example, instead of using actual intersections which appear to have different characteristics or physically modifying a small number of intersections, it will be possible to modify intersection characteristics precisely through simulation to establish the important cues for decisions on turning right as a means of reducing right-turn-against crashes). The next section describes some work done at the Monash University Accident Research Centre (MUARC) to illustrate research applications of simulators.

The stability of driving performance

MUARC is involved in a fundamental young driver research programme investigating parameters of driving performance. This programme utilizes a PC-based driving simulation system originally developed by Systems Technology, Inc. The essence of our approach is summarized below:

- The over-involvement of young drivers in crashes is a large and, to date, intractable road safety problem. A new, more sophisticated approach is both necessary and justified.
- There is a big, but poorly understood, difference between being able to drive (which young drivers generally can do) and being able to drive safely (which young drivers generally cannot do).
- In the past, education and training programmes have invariably failed because they have taught necessary, but not sufficient, skills for safe driving.
- A focus on driving performance is therefore required to identify the necessary and sufficient safe driving skills to be incorporated into education, training and licensing.

At the operational level, the programme's aim has been defined as the identification of differences in driving performance as a function of driving experience. The experimental programme makes the distinction between basic and higher-order driving skills, with the initial work focusing on the basic driving skills of tracking and speed control. Operating on a model of progressive complexity, it is intended that additional tasks in the perceptual and cognitive domains will be singly and concurrently overlaid on the basic driving foundation.

Driving simulation in Australia: opportunities for research and application 253

It has been traditionally thought that the primary problem confronting young drivers is one of roadcraft (analagous to higher-order skills); carcraft or basic driving ability was usually not considered to be of importance as car control skills were obtained through driver training. While the initial outcomes of the investigation of basic driving performance have been formally reported elsewhere (Drummond, Triggs and Schulze, 1992; Triggs and Drummond, 1993), a general conclusion from this work was that the performance of young/inexperienced drivers was less stable, i.e. basic driving performance differences tended to feature in the area of performance variability rather than average performance, with experienced drivers showing more consistent (less variable) performance. As an example, analysis demonstrated that there was a significant experience by trial interaction for lateral position variability ($F(27, 291) = 2.47, p < 0.002$), with the most experienced drivers displaying the greater reduction in standard deviation of lateral position over trials. This result is presented graphically in Figure 24.2.

The graph suggests that experienced drivers demonstrated less variable vehicle speed over trials, although only a marginally significant main effect for speed variability was found on a most/least experienced driver basis ($F(1, 16) = 3.78, p = 0.07$).

It is difficult to interpret the greater variability of basic driving performance of young/inexperienced drivers at this point of the programme. Generally, the differences were small in absolute terms and may simply reflect a less refined driving style with no implications for safe driving. Alternatively, it could be speculated that more variable performance is a symptom of poorer skill development and/or the greater impact of workload on inexperienced drivers (even though the workload level was kept to a minimum).

24.3 Multi-tasking performance

During the driving process, drivers are collecting information, prioritizing tasks, making continuous decisions, implementing actions, monitoring feedback, shedding tasks when driving workload increases and often undertaking a range of non-driving activities. The

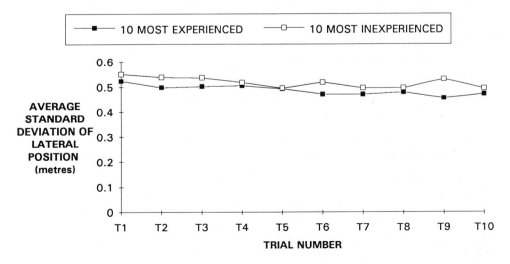

Figure 24.2 Standard deviation of lateral lane position by experience and trial.

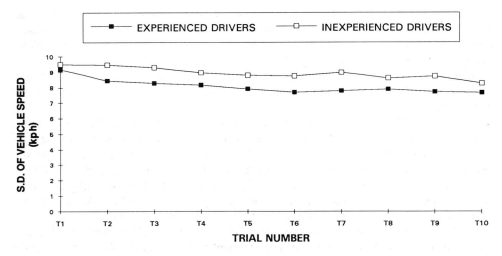

Figure 24.3 Standard deviation of vehicle speed by experience and trial.

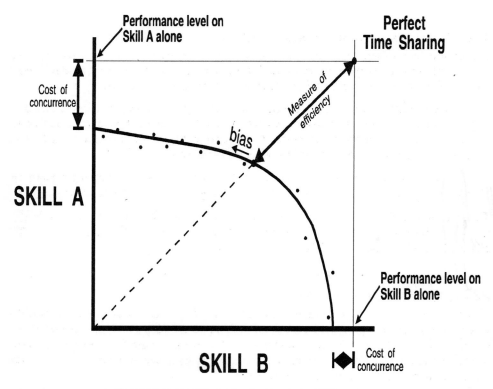

Figure 24.4 Possible dual task scenario for skills A and B.

simultaneous performance of two or more tasks (multi-tasking) is made possible by the allocation of central information processing capacity to each activity. While there are competing theoretical explanations for the way in which central capacity can be distributed, it has been demonstrated that factors influencing this process include the automaticity of the skills, attention level, experience and allocation strategy. Figure 24.4 presents a possible dual task scenario (such plots are often called attention operating characteristics or AOCs).

The use of AOCs – or performance operating characteristics (POCs) as they are called when a combination of disparate tasks is being considered – provides a method for summarizing multi-task performance. It has properties that are akin to its well-known predecessor, receiver operating characteristics (ROCs), in signal detection theory. The AOC approach allows the investigator to estimate the processing costs associated with division of attention or attention sharing in applied tasks, either as the type of task or the characteristics of the individuals involved are changed.

In the context of young drivers, one question that can be addressed is whether the cost of dividing attention is higher for younger drivers (say, 17 years of age) than slightly more mature drivers (say, 19 or 21 years of age). In a task that partially simulates the requirement to make two separate, but overlapping, judgments of vehicle distance, some preliminary evidence suggests that there are systematic and significant differences in the costs of dividing attention across multiple sources of information between 17 and 21 year olds. While there are some methodological challenges involved in applying this method in full-task driving simulation, the method may provide an important basis for exploring deficits in specific aspects of young driver performance.

24.4 Programme applications for simulation

Simulation technology has potential application to many areas of road safety but, because of our young driver orientation, this section focuses on one in particular, driver training. There are two points which need to be emphasized at the very start:

1. Given the current lack of detailed empirical knowledge on appropriate driver training programme content, there would be a strong need for comprehensive evaluation to be built into any such simulator-based programme. Such an evaluation would need to address process issues as well as outcome measures.

2. Simulation brings a number of factors to the training process which have hitherto been unavailable and which may, in themselves, generate positive training benefits.

The first point indicates the need for systematic research effort to optimize simulator-based training programmes – on the reasonable assumption that content issues are also resolved. Several aspects would need to be addressed.

Training, transfer and retention effects

How well does time spent in simulator-based training facilitate driving in the real world? Interestingly, some large airlines are approaching zero flying time (ZFT) training in which the first real-world flight is undertaken on a revenue flight (i.e. with passengers) as part of the flight deck crew. All previous training (for example, to upgrade from one craft to another) is undertaken in a simulator.

Evaluation of simulator-based training raises a number of conceptual and technical issues. A comprehensive evaluation design may have the following features:

Pretest	Train in simulator with feedback	Test on the road	Test retention on the road
Pretest	Train in simulator without feedback	Test on the road	Test retention on the road
Pretest	Train on the road	Test on the road	Test retention on the road

The extent to which such a design could be implemented depends on the nature of outcome variables, the ability to measure such variables in the real world and the available level of experimental control.

Measures of driving performance

The identification of necessary and sufficient measures of driving performance is necessary to evaluate training, transfer and retention effects. Evaluation data can only be viewed with confidence if they are based on valid and appropriate performance measures.

Initial skill acquisition vs long-term skill acquisition

Differences in initial skill acquisition and long-term skill acquisition are important. Relatively little is known about how the driving skill acquired with any training device (in this case, simulation), and demonstrated immediately after training, relates to the skill acquired in the long term.

Training effectiveness

The relative effectiveness of training for specific performance elements versus training aimed at development of higher-order skills is a critical issue because it could have a profound effect on the nature of the training programme.

Uses of simulation

There are many different uses of simulation in training, e.g. driver search, identification, prediction, decision making and execution tasks. It is not known whether acquisition in these different skill categories can be isolated.

The role of the instructor in simulator driver training

Previous research has shown that instructors differ considerably in how they implement instruction. However, there has been little systematic evaluation of how instructor variables influence the acquisition of driving skill.

The development of special purpose part-task simulators
Previous work has shown that such an approach has promise, but more detailed attention is required.

The second point raised at the start of this section related to the benefits that simulator-based training may bestow on driver training. The general consensus on driver training is that, to date, it has failed to improve the safety of novice drivers. While improvements to the content of training programmes will flow from the research effort, there are two aspects to the use of simulator-based training which place it in a potentially favourable position:

1. It is ideally placed to 'import' research outcomes. It is likely that such outcomes will be in the domain of higher order skilled performance and, as such, will require a technological implementation.
2. It imposes a set of processes onto driver training which have either not been available or not available to anywhere near the same degree. These are:
   - training efficiency, i.e. the ability to expose trainees to numerous specific training scenarios, with immediate repetition, in a relatively short time frame;
   - training control, i.e. the ability to specify precisely the type and level of driving task demand;
   - training performance measurement, i.e. the ability to measure precisely and objectively trainee performance, determine rates of improvement and compare trainee performance with average performance of other driver groups (e.g. other trainees, experienced drivers, etc.);
   - training feedback provision, i.e. the ability to provide extensive information on performance, both during and after training; and
   - training environment manipulation, i.e. the ability to generate a risk environment and manipulate this level of risk systematically to enable realistic training tasks to be experienced without the dangers of the real world.

This level of efficiency, control, measurement and feedback has not been previously available and represents a substantial advance in driver training which is consistent with general training principles.

24.5 Conclusion

Rapid advances in simulation technology has meant that high-fidelity simulation technology now has the potential to be incorporated into Australian road safety research and development efforts. This would expand the range of road safety questions which can be addressed by researchers, complement the traditional road safety model and contribute to the development of the next generation of road crash countermeasures.

References

Drummond, A.E., Triggs, T.J. and Schulze, M.T., 1992, Basic driving performance as a function of driving experience, *Proceedings of the 1992 IMAGE VI Conference*, Arizona: IMAGE Society Inc.

Triggs, T.J. and Drummond, A.E. (1993), A young driver research program based on simulation, to be published in Proceedings of the 37th Annual HFES Conference, Human Factors and Ergonomics Society, Santa Monica.

25 Pro-active processing: the hallmark of expertise

Kim Kirsner

25.1 Introduction

The aim of this chapter is to determine the applicability of several models of skill acquisition to expertise in driving. The position advanced is that structural models do not provide a productive platform for research in this area, and that two characteristics, involving dynamic changes in representations, and pro-active processes, respectively, provide a more appropriate platform for theoretical development.

According to Milech, Glencross and Hartley (1989)

> driving is a complex skill involving continuous tracking movements by the hands and arms integrated with feet and leg movements in response to an ever changing, varying and somewhat unpredictable environment.

This definition captures the complexity of the response or motor control side, but the term 'environment' is used to encompass a multitude of data types and input processes. Thus, while the motor control problem remains relatively stable, and can be reduced to a small set of parameters governing the direction and velocity of the vehicle, the perceptual or input side is immensely complex. It includes, for example, not only the immediate perceptual input, but a wide variety of superficially extraneous variables concerning the impact of visibility on perceptual processing, the impact of weather conditions on vehicle performance, and conditional probabilities associated with an almost infinite number of possible objects, many of which are or could become hazards. Furthermore, while an expert driver's understanding and mastery of these variables will be based on 'perceptual-motor' experience in some cases, in many other cases it must be based on knowledge rather than experience, a point that is underlined by the proposition that accidents involve rare if not unique combinations of variables. The implication of this line of argument is that structural distinctions based on the proposition that different types of knowledge are represented in different systems (Rasmussen, 1986) must be set aside in favour of process models that emphasize ways in which different classes of information can be brought together to support a mental model of the driving environment – a mental model that drivers can monitor and refine in order to achieve pro-active or anticipatory control over their behaviour on the road.

Analysis of the literature on expertise suggests that the major differences between the perceptual performance of novice and expert drivers involves three factors:

1. the shape of the perceptual envelope – that is, distribution of attention in the evolving region that could be monitored to ensure vehicle safety;

2. the extent to which, and the rate at which, relevant information can be extracted from the perceptual envelope; and

3. the assignment of risk values to actual and potential hazards in the envelope.

Where attention is concerned, the research indicates that novices do not search selectively the most appropriate portions of the driving environment, and that they tend to concentrate on areas closer to and in front of the vehicle (Kundel, Nodine and Carmody, 1978; Mourant and Rockwell, 1972). Their attention is distributed more evenly and therefore less selectively among the available options. How does experience modify this skill? If psychological models of skill acquisition are accepted as a guide, the obvious assumption is that experience including feedback about performance will be critical. However, while experience can be expected to play a role – as novices are forced to respond to objects and events that emerge from beyond the projected vehicle envelope – it is also likely that their perceptual envelopes are shaped by tacit knowledge that is shared among experts and novices about driver safety and performance.

Where perceptual analysis is concerned, there is consistent evidence that experts develop more and more functional schema, and that this change confers advantages on them in regard to processing time. Thus, for example, the early work by Chase and Simon (1973) indicated that experts included many more chess pieces in each pattern during reconstruction than novices. Similar results have been reported for radiological and geological problems (Hawkins, 1983; Lesgold, 1984), as if the granularity of representations rather than the number of representations is transformed by the development of expertise. Perhaps the most important implication of this change is that expert viewers can extract more information per unit of time from a scene than novices, enabling them to handle routine perceptual processing with a modest proportion of their mental capacity while reserving a more substantial proportion for non-routine events. However, here too there is a paradox. As changes in traffic control, road design and vehicle performance have eliminated or reduced many forms of accidents, they have reduced the extent to which individuals can acquire direct experience of those events that precede and accompany accidents, and, therefore, the means of acquiring expertise by direct experience. Thus practised drivers are experts at routine problem solving, but they are not in a position to become experts at managing non-routine problems on the road because such problems are, by definition, rare. A safe driver must therefore acquire or develop his or her expertise in another way. The specific hypothesis advanced in this essay is that this mode of learning relies on the use and practice of mental models, perceptual-motor plans that people implement to prepare themselves for specific types of actions; actions that might be required if the anticipated set of circumstances is experienced. This hypothesis receives weak support from evidence that people can use predictive information under both laboratory and driving conditions (Downing, 1986; Shinar, 1985). However this research does not address the fundamental question concerning the long-term impact of practice involving reference to mental models, potential cues and hypothetical scenarios on the perceptual fluency of the driver.

Where risk assessment is concerned, the role of mental intervention cannot be doubted. People do not acquire sufficient experience about the risks associated with decisions on the road to build an actuarial map of the driving world, a point conceded by Hoyos (1988). They can only acquire this knowledge vicariously, and, if they are to take advantage of this source of information, they can only do so by integrating it into their mental models of on-road behaviour prior to the critical event.

In summary, the central proposition under review here is that a substantial proportion of an expert driver's expertise is acquired vicariously, under off-road conditions, and that this knowledge can be instantiated – as expertise – if the student develops and practises mental models that bring this information into functional contact with output plans that can be used for vehicle control. We have suggested, furthermore, that this proposition applies to many aspects of expertise, including specifically scanning strategies, perceptual analysis and perceptual fluency, and risk assessment.

25.2 Dynamic processes

This chapter is concerned with the control of dynamic processes. It is a basic tenet of the author's approach to research in this area that it should be possible to explain a very wide range of phenomena by reference to a small set of basic principles, and that domain-specificity is to be avoided, if possible. Therefore a broad approach is used here, including illustrations and arguments based on a variety of models and paradigms. The distinction between static and dynamic tasks, involving lexical problems and industrial process control for example, may not even be honoured in the twenty first century models.

As young sciences, it is not surprising that the cognate disciplines of cognitive psychology and human factors give prominence to distinctions between different types of expertise. In the area of memory, for example, distinctions between concepts such as short-term memory and long-term memory, episodic memory, and semantic memory, and explicit and implicit memory have dominated research for three decades (Broadbent, 1958; Waugh and Norman, 1965; Tulving, 1972; Tulving, 1983; Schacter, 1987; Roediger, 1990). A similar trend may be discerned in recent work in human factors, where Rasmussen (1983), for example, identified three modes of processing information:

- skill-based behaviour,
- rule-based behaviour, and
- knowledge-based behaviour.

Where driving is concerned, the modes may be illustrated by reference to the way in which drivers manage the distances between their own vehicle and that of a driver in front of them. If this is achieved by, for example, monitoring and maintaining the angle subtended by the width of the vehicle under observation, a taxing tracking task, skill-based behaviour is involved. If, however, distance is controlled by a rule such as

> IF I can read the number plate of the vehicle in front of me
>
> THEN I am too close to it

rule-based behaviour is involved. The operation of knowledge-based behaviour involves reference abstract knowledge about the system or environment, including for example, knowledge about the formal structures of relationships and processes. One example, rarely encountered in Western Australia, might involve concern about the presence of 'black ice'. In this case, driver behaviour might be modified by knowledge about:

- the conditions under which black ice is found, and
- the effect of black ice on adhesion.

25.3 Rasmussen's model

Vicente and Rasmussen (1992) reiterated the basic tenets of this taxonomy, and characterized it in the following terms.

1. Skill-, rule-, and knowledge-based behaviour involve mutually exclusive information processing modes.

2. Although these modes involve qualitatively different principles of analysis and representation, selection is determined by the way in which information is interpreted by

the user. Selection is not determined by intrinsic stimulus properties, but by the 'reading' selected by the user. A single display can therefore be 'read' or exploited by all three modes of processing.

3. The three modes treat stimuli as signals, signs and symbols, respectively. According to Rasmussen (1986: 108), signals are:

> ... sensory data representing time-space variables from a spatial configuration in the environment, and they can be processed by the organism as continuous variables.

Signs, by contrast:

> ... indicate a state in the environment with reference to certain conventions for acts. Signs are related to certain features in the environment and the connected conditions for action. Signs cannot be processed directly; they serve to activate stored patterns of behaviour.

Symbols represent:

> ... other information, variables, relations, and properties, and can be formally processed.

These modes are illustrated in Rasmussen (1986, Figure 9.3).

4. The three modes support three distinct levels of cognitive control. This aspect of Rasmussen's account is illustrated in Rasmussen (1983, Figure 1). They involve skill-based, rule-based and knowledge-based behaviour, respectively. The position advanced here is that Rasmussen's model assigns too much emphasis to the static and structural characteristics of expertise, and insufficient emphasis to problems associated with transformation and change, and with the way in which information from the identified structures is used to control behaviour.

It should be noted however that Rasmussen (1983) anticipated some of the issues raised in the following sections of this chapter, but I will return to this issue. The structural approach is open to challenge for several reasons. The first of these stems from the indeterminate nature of research involving fractionation in cognitive psychology. If questions about the inferential status of double dissociation are set aside (Dunn and Kirsner, 1988), it is apparent that the experimental and the neuropsychological applications of the procedure tend to support proliferation of hypothetical memory systems. At the extreme, evidence of dissociation may be obtained for all tasks, all stimulus classes, and even all stimuli. Put in other words, the technique, if valid, may be too powerful. It offers analysis without synthesis. An extension of this approach to the domain of expertise may not be productive.

The second question concerns the significance of the more or less ubiquitous power law (Anderson, 1982; Rosenbloom and Newell, 1983; Newell, 1991). The power law describes the relationship between practice and performance (i.e. reaction time, or fluency) on many tasks. This function describes the relationship for a variety of laboratory, industrial and other activities. For example, it has been observed for tasks as disparate as cigar rolling, lexical decision, syllogistic reasoning, and even book-writing. It has also been found for tasks which, strictly speaking, do not involve speeded performance, elective inspection time on premises prior to syllogistic decision (Speelman, 1991). More surprisingly perhaps, it has also been observed for tasks which involve organizational performance, including for example, the relationship between unit production (i.e. first truck, second truck, third truck, etc.) and per unit production time, in hours per unit (Argote and Epple, 1991). That is, time to build trucks is a power function of unit number.

The critical point for the present argument is that there are few hints of discontinuity in the relevant literature (Nerb, Krems and Ritter, 1993). Even though some models include specific provision for the operation of qualitatively distinct processes at different stages of skill acquisition (Anderson, 1982; Logan, 1988), the typical pattern of results suggests that the same principle if not the same processes can be invoked to explain performance changes during all stages of skill acquisition. If knowledge-, rule- and skill-based reasoning involve distinct processes, why does practice on tasks which arguably straddle this range conform to the same principle? It should be noted of course that these data do not falsify the proposition that different processes are involved at each level; the presence of continuity in the fluency data does however require an answer from researchers who table theories which focus on process or system discontinuity.

In a recent refinement of Anderson's approach, Kirsner and Speelman (1993) have suggested that even a relatively simple, static task such as lexical decision depends on RT components which, although they are at very different positions on the practice function, nevertheless reflect the power law. It may be noted that the components in question in lexical decision have also been attributed to qualitatively different stages or processes by other authors, involving data-driven and conceptually driven processes, or perceptual and episodic processes. The point is then that authors who endorse multi-system or multi-process explanations of performance must account for the challenge posed by evidence that the power law influences all of the performance components in the relevant data.

Argument based on the power law is not without its problems however. Perhaps the most vexing of these stem from the fact that fluency (i.e. RT) is by no means the only parameter of interest. The power law has not yet been refined to provide a quantitative explanation of changes in other parameters of skill; involving:

- mastery, i.e. changes in accuracy during early stages of practice,

- precision, i.e. in tasks which require precise movements, or

- vulnerability, i.e. the extent to which skill is vulnerable to stress or other extraneous variables in particular.

One final problem stems from evidence that context can override practice under many circumstances. For example, in language comprehension, when contextual cues are available, the amount of variability in the data explained by word frequency usually declines dramatically.

The third question involves the use of the term 'behaviour' to characterize the role of skills, rules and knowledge. The proposition that information can be extracted from control displays in qualitatively different ways is not in dispute here. The critical issue concerns the way in which these putative information processing modes are reflected in behaviour. One possibility, suggested by Rasmussen (1986, Figure 9.2), is that information from the rule-based and knowledge-based domains is translated into the language or code of the skill-based domain, so that it enters 'behaviour' indirectly.

Another possibility is that each type of knowledge or level of control is responsible for its own motor planning functions. However, the first and second approaches do not explain performance if the relationship between the various levels of control is conditional, and non-linear. Consider the following problem involving skidding (on black ice or gravel) and camber. Assume, furthermore, that the controlled responses to these variables involve knowledge-based behaviour and skill-based behaviour respectively. There is no problem if the controlled responses to variation in these variables are additive. But if the sign or even just the magnitude of the controlled response to skidding is contingent on the current state of the controlled response to camber, it follows that accounts which involve either

- uni-directional translation from the knowledge-based level to the skill-based level, or
- independent and additive motor planning contributions from the rule-based and skill-based levels,

may be inadequate. The missing elements involve intentions and integration. If the components identified above involve distinct processes, it may be necessary to assume that these processes communicate intentions as distinct from commands to an additional stage, where they are prioritized, and, if necessary, integrated.

A similar problem may be identified in what is one of the oldest skills, speech. The term prosody is used to refer to the way in which fundamental frequency, amplitude and the duration of segmental and supra-segmental speech are used to convey information about communicative intent (e.g. about questions, statements, and topic change), psycho-linguistic status and emotion. There is evidence to suggest that the mapping relationship between the communication category and acoustic feature does not involve invariant features. Instead, the data suggest that the way in which information about each of these state-space attributes is communicated is sensitive to the status of the entire, multi-variate, speech system, where this includes phonation, respiration and vocalization. For example, there is evidence from spontaneous speech that when people signal topic changes, they may achieve this by either:

- an increase in fundamental frequency, or
- an increase in amplitude, where selection may be moderated by the current status of psycholinguistic and even physiological variables (Hird, 1994).

In other words, speakers are able to use a wide variety of mechanisms to achieve communication goals, the chosen mechanism for a given purpose will be constrained by the way in which those mechanisms are being used for other purposes at the critical moment, selection will be restricted accordingly, and a given goal may therefore be achieved via different mechanisms on different occasions. This analysis suggests that far from being additive, the mechanisms selected to achieve these goals must be selected conditionally, and that integration must therefore precede motor planning. The process therefore involves integration, the solution to each communicative problem depends on the solution to other communicative problems, and the entire process must therefore involve parallel, interactive processing as distinct from serial, additive processing.

The implications of this analysis are illustrated in Figure 25.1. Three modifications from the original by Rasmussen (1983, Figure 1) merit consideration:

1. The three domains are retained, although their role is restricted to stimulus analysis and, possibly, representation.
2. The assumptions that
    - knowledge-based planning is converted into rules, and

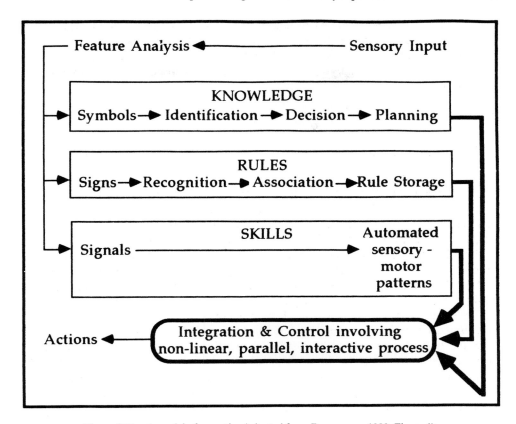

Figure 25.1 A model of expertise (adapted from Rasmussen, 1983, Figure 1).

- rules are converted into automatic sensory-motor patterns

  are discarded in favour of the proposition that a separate stage is required to integrate information from the three modes, and that this must occur before motor planning and therefore action is implemented.

3. The final modification involves the character of the additional stage, concerned with integration and control. Characterization of this stage involves reference to non-linear, parallel, interactive processes.

The implications of the above analysis for driving are clear. Consider the problems encountered by vehicle control on a wet road where black ice has not yet been seen by the driver, although it has been foreshadowed in weather reports. Information about the impact of black ice on road-holding could be stored in the form knowledge and/or rules, but this information must be integrated with direct visual feedback about moment-to-moment vehicle performance. The proposition advanced here is that the process of integration is critical; information about mode of reception and representation is of little consequence. It seems probable, furthermore, that a project designed to reveal which aspects of motor control are moderated by skill-based behaviour and knowledge-based behaviour may not be productive.

In summary then, the criticism of Rasmussen's model is that it emphasizes the static properties of representation, concerning intrinsic differences between different types of expertise rather than the dynamic transformations which are, in this author's opinion, central to the acquisition of expertise. While the differences described by Rasmussen may be obvious at the input level, as a natural extension of the proposition that different types of information are available in the environment, and these differences may be represented in the mind in the attributed form, as knowledge, rules and skills, surely the critical issues for research into expertise involve the processes which transform these representations about various forms of expertise into behaviour.?

25.4 Mental models

Does the notion of a mental model provide a more valuable metaphor for the analysis of dynamic processes? In order to consider this problem, we will describe what we mean by a mental model, and then apply it to dynamic problems. For the purpose of the arguments presented here, consideration is restricted to a specific form of mental model, i.e. to user's mental model of a physical or conceptual system; no consideration will be given to the designer's mental model of the system, or to the designer's mental model of the user.

Wilson and Rutherford (1989) provide a useful definition for the present purpose:

> A mental model is a representation formed by a user of a system and/or task, based on previous experience as well as current observation, which provides most if not all of their subsequent system understanding and consequently dictates the level of task performance.

A number of supplementary propositions merit consideration, however. The most important of these involves the concept of prediction, and the concomitant assumption that a mental model can be 'run' or 'run off' in some sense (Norman, 1983; Rouse and Morris, 1986).

The crucial element in the above account is that a mental model can be used to predict the way in which a system will behave under conditions not yet experienced by the holder. Mental models can therefore be used by people to understand and anticipate the behaviour of physical systems. A second proposition, also introduced by Norman (1983), is that mental models are naturally evolving models. This proposition is given additional consideration below. The third proposition is that mental models can be formed at many levels of abstraction, ranging for example from the procedures used to test system readiness (e.g. pre-dive operations in a submarine) to the static architecture of the system (e.g. the organization of the electronics or hydraulic systems of a submarine). As suggested by Wilson and Rutherford (1989), each form of description may involve a unique mental model for a given system, and no one person can be expected to carry all of the possible or even necessary mental models for the operation of a given system. The final supplementary proposition to be considered here is that a mental model consists of the set of schemata instantiated at a particular moment; the obvious implication being that it is a sample from a far larger set of stored schemata, the balance of which remain uninstantiated during each and every operation (Brewer, 1987; Rumelhart, 1984).

25.5 Descriptions and records

The model depicted in Figure 25.2 was designed to capture some of the ideas introduced above. It is based on concepts developed by Norman and Bobrow (1979),

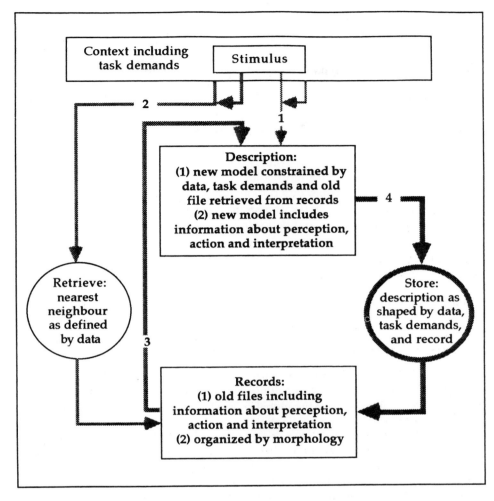

Figure 25.2 Descriptions and records: the assumption that detailed records are preserved about the processes invoked to solve problems.

Treisman (1984), and Kirsner and Dunn (1985). The model includes four data transactions. The first of these involves data from:

- the stimulus, and
- context and task demands (#1), explicit or otherwise.

Information from these sources is supplied to working memory where a composite 'description' of the current event or episode is formed (Norman and Bobrow, 1979). While the description is constrained directly by the current stimulus, it is also shaped by context and task demands. Context, for example, could determine whether or not a lexical stimulus such as 'bank' should be interpreted in financial or watery terms.

The second transaction is also triggered by the stimulus, context and task demands (#2). But this transaction involves a 'retrieve' command to long-term memory, the content of which is a set of records. The critical assumption is that a record is preserved about each and every instance or event. It is further assumed that these records are:

- organized in terms of morphologically defined clusters – where morphology is defined in structural terms which transcend modality, and which can, therefore, be applied to objects and scenarios as well as linguistic and textual material, and
- retrieved via a nearest-neighbour selection process which operates on structural or morphological data.

The third transaction involves the record or file which, when retrieved from long-term memory (#3), is supplied to the working memory to provide a scaffold for the formation of a description of the current event. It should be noted that because this file has been selected by reference to the morphology of the stimulus, it also defines the return address to be used when the new description, formed by the first three transactions, is deposited in long-term memory (i.e. transaction #4). In other words, descriptions are prepared in working memory, and deposited as records in long-term memory when composition is complete.

Four other issues merit consideration. First, because used descriptions are deposited as records on existing, morphologically defined pegs, practice effects will accumulate across morphologically defined instances, thereby providing a basis for practice and, by extension, power law effects. Definition of a mechanism to explain the ubiquitous power law relationship between practice and fluency awaits further exploration however.

Second, because contextual information contributes to record selection, the level of granularity of the record which is recruited from, and subsequently deposited in long-term memory will be context-sensitive. Thus, the word 'bank' may be stored on different morphological pegs depending on either the local context (e.g. as a lexical term concerning water or money) or the global context (e.g. as a lexical term, or as part of an argument about banking regulation).

Third, one of the features of our initial account (Kirsner and Dunn, 1985) was that it included a detailed perceptual record of the stimulus, where a record might include information about language (e.g. English versus French), mode (picture versus word), modality (spoken versus printed), case (i.e. for printed words), speaker's voice (i.e. for spoken words), and other stimulus attributes. More recently (Kirsner and Speelman, 1993), this author has argued that it may be necessary to include records about output or production operations as well as input or perceptual operations.

Fourth, are separate metaphors required for static and dynamic processes? Where should we draw the line between static and dynamic processes? Consider, for example, the relationship between a raconteur and his or her audience. The account offered by the raconteur is nothing if not dynamic. It involves a sequence of events which unravel over time, and space. Presumably, therefore, comprehension and reproduction involve the retention and retrieval of information about events which are distributed over space and time. Presumably, therefore, the representation must carry information about the spatial and temporal elements in the story. Which features, if any, distinguish the analysis and representation of information for this story from the analysis and representation of some dynamic industrial process?

25.6 Descriptions, records, and mental models

The proposition advanced here is that a mental model is but a specific form of description and that, like descriptions, a detailed record is preserved in long-term memory, for use during subsequent perceptual operations. The only feature which distinguishes a mental model from other descriptions, and records, is that it can be 'run' to enable prediction about the behaviour of events and systems which the user has not experienced.

The arguments advanced above suggest that it may be necessary to distinguish between reactive and pro-active processes where mental models are concerned. Reactive processes are available to inexperienced drivers or operators, who, when faced with a system management problem, can manipulate control devices to reduce discrepancies between the observed and desired behaviour. The extreme novice does not even need to refer to an old record to guide his or her behaviour; mere trial and error should reduce the discrepancy between the observed and desired states without practice. It may be assumed however that practice will provide the system operator with records which can be reactivated to calibrate control decisions with the magnitude of anticipated changes in the physical system. The experienced operator is in a position to 'run' his or her mental model in order to behave pro-actively, i.e. to change the physical system to reduce the magnitude of the discrepancy which will arise when the physical system undergoes some anticipated change.

The distinction between reactive and pro-active performance provides an interesting point of contact with the distinction between skills, rules and knowledge. A novice who is restricted to reactive processes cannot, by definition, use knowledge to moderate his or her performance. They do not possess and they cannot therefore use information about the calibration between system changes and behavioural control. An expert, on the other hand, is in a different position. If experts can translate their knowledge into the metric of the integration stage, they are in a position to activate a pro-active process given signals, signs or symbolic data.

25.7 An experimental illustration

Aspects of the following study were described in detail by Speelman et al. (1993). Figure 25.3 depicts part of one of the displays used to manage the distribution of power throughout an electricity network. The display was derived from work in progress on development of a training system for the Perth Metropolitan Electricity Grid (Milech et al., 1993). The figure depicts a generator which feeds power to substations, each of which has a network of transformers and circuit-breakers. The controllers manipulate the transformers and circuit-breakers to ensure that the power requirements are met.

Figure 25.3 is a simulation of a small section of a grid. The analogue display was based on knowledge elicited from expert power supply controllers. According to their verbal reports (Boase-Jelinek, personal communication), the experts represented voltage readings along the line as an object that changed shape as the voltage changed. This object was often described as a 'bending bar'. As the amount of power being transferred by the line increased, the voltages change, and the bar changes shape accordingly. The operator's aim is, then, to keep the bar straight and within prescribed limits by appropriate manipulation of the controls. The bending bar depicted in Figure 25.3 is, therefore, a visual depiction of a concept described by expert power supply controllers (Speelman et al., 1993).

A simplified version is displayed in Figure 25.4. This version consists of just two substations and one power source. The bending bar is being used here to provide explicit feedback about system states. The bar indicates the stability of the system. If the power going in and out of the system is not balanced, the bar will bend. The subject's task is to manipulate the transformers in order to keep the bar in a straight line and, thereby,

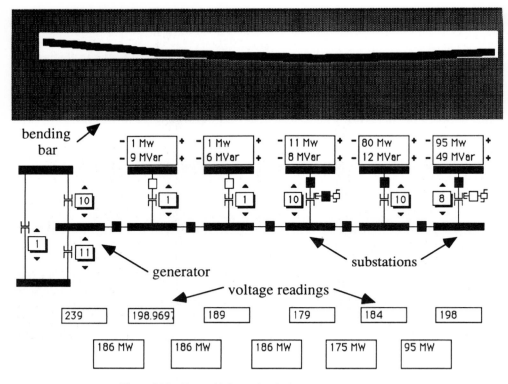

Figure 25.3 Type of information included in SECWA grid.

maintain system stability. The time structure of the task compresses a day's worth of voltage variation into one minute. These simplifications ensured that the task was within our subject's capabilities. The line load changed every second. If the subject does not adjust one of the transformers, the 'bar' will change shape accordingly. The algorithm which relates changes in line load to changes in the shape of the bar is based on one that controls the original system. The subjects' task is to adjust the transformers to keep the bar straight within the limits defined in the display. Subjects are also informed that the system may be damaged when it is unstable, and given points based on their performance.

Simplification of the power system has however changed the nature of the operator's task. Instead of processing a number of parameter values (e.g. voltage values, power values, transformer settings) in order to decide upon a course of action, our subject-operators must strive for homeostasis between the generation and consumption of power by manipulating transformer settings. The task is now a closed-loop system where deviations in the shape of the bar can be corrected as they occur. Psychology subjects, unschooled in electrical engineering, have demonstrated that this version of the task is simple to learn. We have not, however, tried to determine the extent to which the improved 'learnability' of our miniature power supply control system reflects task simplification *per se*, or provision of a mental model based on expert knowledge.

As described above, subjects enjoyed just one strategic option. They were not provided with advance information about system behaviour, and they were therefore restricted to a 'reactive' strategy, i.e. they could correct system instability when, or after,

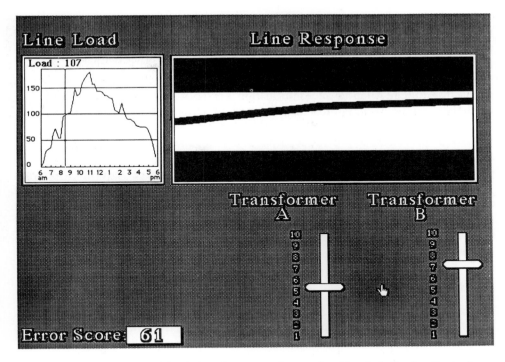

Figure 25.4 Experimental simulation of power supply control problem including 'bending bar'.

it was evident from the shape of the bending bar. In practice however, we sometimes provided subjects with additional information about future changes in the load; see the window in Figure 25.4. Our subjects could, if they chose, use this information to anticipate and therefore avoid changes in the shape of the bar, i.e. instability. In other words, we placed our subjects in a situation where

- they acquired a mental model about the operating characteristics of a system, and
- they were provided with advance warning about system changes, which, provided that they 'ran' their mental model correctly,
- they could use to improve system performance.

It should be noted, in addition, that the task involves several dependent variables or parameters. One parameter involves calibration between the magnitude of system changes and the magnitude of the subject's responses. This parameter can of course be calculated under both reactive and pro-active conditions, i.e. when the subject is restricted to reactive processes by the display procedure, and the subject is provided with data which he/she can use to anticipate and avoid system instability. The second parameter involves timing. Under reactive conditions the timing parameter is, in effect, reaction time. But under pro-active conditions, the distributions of subject responses should gradually advance in time, until they precede and therefore anticipate changes in system performance. They may also become more concentrated, i.e. leptokurtic, as expertise is acquired.

The measures which we have isolated indicate that our subjects progress from an initial stage during which they use an exclusively retroactive strategy to a second stage during which they use both retroactive and pro-active strategies. At the beginning of the experimental session, the magnitude of subject's responses are strongly correlated with the magnitude of the last voltage change, as if they are using a retroactive process. During this stage they are unable to use information that is available in advance about some voltage changes. After several days of practice however, they not only use a retroactive strategy, when advance information is not available, but a pro-active strategy, when they receive advance information about voltage changes. In this initial study, we did not establish separate measures for calibration and timing, but we have subsequently refined the procedure to provide separate indices for these variables. The new procedure will enable us to explore the way in which these two variables are influenced by practice and other variables. We will be able to compare the acquisition of what Rasmussen would refer to as skill- and knowledge-based behaviour, involving signals and symbols, respectively. The new study will also enable us to operationalize pro-active processing, comparing the rates at which reactive and pro-active processes develop, and generally explore the use of dynamic mental models under the influence of stressors and other variables.

25.8 Conclusion

Two links with driver expertise merit consideration. The first may be discussed by reference to a hypothetical envelope which drivers deploy in front of, and to the sides of their vehicle's projected path. Whereas the expert may construct and maintain a detailed mental model of the unfolding environment, including risk analyses where information is unavailable (e.g. for the probability that a dog will emerge from behind a masked position), this author assumes that the novice driver will enjoy access to a much smaller and less articulated envelope. The implication for people with powerful simulators is that they should study not only how novices and experts respond to rapidly changing events, but also to the way in which novices and experts use knowledge as distinct from practice to prepare themselves for non-routine but high risk scenarios.

The second link with driver expertise involves practice. Whereas extensive practice with chess and radiology problems will automatically expose the novice to non-routine as well as routine problems, extensive practice on the road might provide no exposure to non-routine problems at all, or, if it does, the result may be catastrophic. The implication of this problem is that it is necessary to understand processes that transform knowledge from an explicit form into an implicit operational procedure that will protect relatively inexperienced drivers when non-routine problems are encountered. It is suggested that this orientation will require reference to dynamic as distinct from structural models of skill acquisition.

In conclusion, perhaps we have used Rasmussen too harshly? For example, Rasmussen (1986: 138) entertained the proposition that problem solving can be transferred from knowledge-based behaviour to rule- or even skill-based behaviour: 'If it is possible to reinterpret the symbolic representations at another level of abstraction at which rules for manipulation are available.'

References

Anderson, J.R., 1982, Acquisition of cognitive skill, *Psychological Review*, **89**, 369–406.
Anzai, Y., 1984, Cognitive control of real-time event-driven systems, *Cognitive Science*, **8**, 221–54.
Argote, L. and Epple, D., 1990, Learning curves in manufacturing, *Science*, **247**, 920–82.

Brewer, W.F., 1987, Schemas versus mental models in human memory, in Morris, P. (Ed.), *Modelling Cognition*, Chichester, England: Wiley, pp. 187–97.
Broadbent, D.E., 1958, *Perception and Communication*, New York: Pergamon.
Chase, W.S. and Simon, H.A., 1973, Perception in chess, *Cognitive Psychology*, 4, 55–81.
Donges, E., 1978, A two-level model of driver steering behaviour, *Human Factors*, 20, 691–707.
Downing, C.J., 1986, Expectancy and visual-spatial attention: Effects on perceptual quality, *Journal of Experimental Psychology: Human Perception and Performance*, 14, 188–202.
Dunn, J.C. and Kirsner, K., 1988, Discovering functionally independent processes: the principle of reverse associations, *Psychological Review*, 95, 91–101.
Hawkins, D., 1983, An analysis of expert thinking, *International Journal of Man-Machine Studies*, 18, 1–47.
Hird, K., 1994, *Communicative Intent: The Role of Declination and Resetting*, Doctoral dissertation, University of Western Australia.
Hoyos, C.G., 1988, Mental load and risk in traffic behaviour, *Ergonomics*, 29, 377–91.
Kirsner, K. and Dunn, J.C., 1985, The perceptual record: a common factor in repetition priming and attribute retention, in Posner, M.I. and Marin, O.S.M. (Eds), *Mechanisms of Attention: Attention and Performance XI*, Hillsdale, New Jersey: Lawrence Erlbaum.
Kirsner, K. and Speelman, C., 1993, *Is Implicit Memory necessary?* Paper presented to the 36th Annual Conference of the Psychonomics Society, Washington, D.C.
Kragt, H. and Landeweerd, J.A., 1973, Mental skills in process control, in Edwards, E. and Lees, F.P. (Eds), *The Human Operator in Process Control*, London: Taylor & Francis.
Kundel, H.L., Nodine, C.F. and Carmody, D., 1978, Visual scanning, pattern recognition and decision-making in pulmonary nodule detection, *Investigative Radiology*, 12, 175–81.
Lesgold, A.M., 1984, Acquiring expertise, in Anderson, J.R. and Kosslyn, S.M. (Eds), *Tutorials in Learning and Memory*, New York: W.H. Freeman & Co.
Logan, G.D., 1988, Automaticity, Resources and Memory: Theoretical Controversies and Practical Implications, *Human Factors*, 30, 583–98.
Milech, D., Glencross, D. and Hartley, L., 1989, Skills acquisition by young drivers: Perceiving, Interpreting and responding to the driving environment, Report submitted to the Federal Office of Road Safety, Canberra.
Milech, D., Waters, B., Noel, S., Roy, G. and Kirsner, K., 1993, Student modelling in hybrid training systems, *Proceedings of HCI International 1993 (the 5th International Conference on Human-Computer Interaction and the 9th Symposium on Human Interface (Japan)*, Orlando, Florida.
Mourant, R.R. and Rockwell, T.H., 1972, Strategies of visual search by novice and experienced drivers, *Human Factors*, 14, 325–35.
Nerb, J., Krems, J.F. and Ritter, F.E., 1993, Rule learning and the power law: A computational model and empirical results, *Proceedings of the Fifteenth Annual Meeting of the Cognitive Science Society*, Boulder, Co.
Newell, K.M., 1991, Motor Skill Acquisition, *Annual Review of Psychology*, 42, 213–37.
Norman D., 1983, Some observations on mental models, in Gentner, D. and Stevens, A.L. (Eds), *Mental Models*, Hillsdale, NJ: Lawrence Erlbaum.
Norman, D.A. and Bobrow, D.G., 1979, Descriptions: an intermediate stage in memory retrieval. *Cognitive Psychology*, 11, 107–23.
Rasmussen, J., 1983, Skills, rules, knowledge: Signals, signs and symbols and other distinctions in human performance models, *IEEE Transactions: Systems, Man and Cybernetics*, 13, 257–67.
Rasmussen, J., 1986, *Information processing and human-machine interaction: An approach to Cognitive Engineering*, New York: North Holland.
Roediger, H.L., 1990, Implicit memory: retention without remembering, *American Psychologist*, 43(9), 1043–55.
Rosenbloom, P.S. and Newell, A., 1983, The chunking of goal hierarchies: a generalized model of practice, in Michalski, R.S., Carbonell, J.G. and Mitchel, T.M., (Eds), *Machine Learning: An Artificial Intelligence Approach*, Vol II, Morgan Kaufmann.: Los Altos.
Rouse, W.B. and Morris, N.M., 1986, On looking into the black box: Prospects and limits in the search for mental models, *Psychological Bulletin*, 100, 349–63.
Rumelhart, D.E., 1984, Understanding understanding, in Flood, J. (Ed.), *Understanding reading comprehension*, Newark, NY: International Reading Association, pp. 1–20.

Schacter, D.L., 1987, Implicit memory: history and current status, *Journal of Experimental Psychology: Language, Memory & Cognition*, **13**(3), 501–18.

Shinar, D., 1985, The effects of expectency, clothing reflectance and detection criterion on night-time pedestrian visibility, *Human Factors*, **27**, 327–33.

Speelman, C., 1991, *The effect of transfer on the shape of learning functions*.

Speelman, C., Boase-Jelinek, D. and Kirsner, K., 1993, The role of mental models in complex dynamic environments, *Proceedings of HCI International 1993 (the 5th International Conference on Human–Computer Interaction and the 9th Symposium on Human Interface (Japan)*, Orlando, Florida.

Tulving, E., 1972, Episodic and semantic memory, in Tulving, E. and Donaldson, W. (Eds), *The Organization of Memory*, New York: Academic Press.

Tulving, E., 1983, *Elements of Episodic Memory*, Oxford: Clarendon Press.

Vicente, K.J. and Rasmussen, J., 1992, Ecological interface design: Theoretical foundations, *IEEE Transactions on Systems, Man and Cybernetics*, **22**, 590–606.

Waugh, N.C. and Norman, D.A., 1965, Primary memory, *Psychological Review*, **72**, 89–104.

Wilson, J.R. and Rutherford, A., 1989, Mental models: Theory and application in human factors, *Human Factors*, **31**, 617–34.

Index

aboriginal population 51, 53, 56–7
accidents
  age of driver 7–8, 12, 41, 191, 249–50, 252
    NSW 98–9, 103
    Western Australia 51–7, 60–1, 85–6
  alcohol 57, 87, 92–3, 101, 104, 189, 233
  alertness 219
  animal collisions 51, 54, 62
  avoiding collisions 12, 45–6, 216
  CMVs 33, 133
  driver behaviour 4, 181, 185, 233, 237–46
  driver's hours 3, 6, 9, 41–6, 97, 100–4, 133, 219, 226
  drugs 67–72
  expertise 260
  fatal 59–66, 80–1
  fatigue-related 4, 7–12, 97–105, 189
  freight 18
  head-on 53–4, 62, 65
  livestock transport 22
  methodological research 155–6, 159, 161–4
  night-time 7–8, 42–3, 51–4, 88, 92–3
  rear-end 8, 51, 53–4, 57, 62, 216
    deceleration indicators 107–8, 110, 112, 117
  region in WA 8, 51–7, 73–86
  road design 63, 87–94
  simulators 136–9, 141–4, 146–7, 216
  time of day 6, 42–4, 46, 60–1, 84, 98–104,
    see also run-off-road accidents; single-vehicle accidents
additive factor logic 167–79
advanced transport telematics (ATT) 10–11, 161, 164, 186
age of driver 11, 152, 154, 181
  accidents 7–8, 12, 41, 191, 249–50, 252
    NSW 98–9, 103
    Western Australia 51–7, 60–1, 85–6
  alcohol 191, 193–202
  CMVs 37–8
  driving behaviour 3, 11–12, 233, 237
  elderly 252
  mediating fatigue 157, 163, 165
  simulators 13, 135, 249, 253–7
aggressive–competitive behaviour 234, 239–42, 244, 246
alcohol 7, 11, 18–19, 46, 97, 152, 181–6

accidents 57, 87, 92–3, 101, 104, 189, 233
  additive factor 167–79
  driving and enforcement 67–72
  education programme 8, 121–5
  and fatigue 8, 30, 133, 189–205
  vision 63–5
alertness 11
  alcohol and fatigue 199, 201
  CMV drivers 33–6, 38
  driver's hours 45–6, 219, 224, 226
  monitoring system 219–29
  steering 45, 222, 225–6
attention operating characteristics (AOCs) 255
audio stimulation 38, 62, 127–32
  simulators 207–10, 213, 216
Australia 3–13, 42, 44, 127, 249–57
  accidents 8–9, 42, 44, 51–7, 59–66, 73–86, 97–105, 189
  bus drivers 121–5
  driver behaviour 233–47
  freight 15–20
  livestock 21–3
  Western Australia 8–9, 15–20, 21–3, 51–7, 59–66, 67–72, 73–86, 151–4

black ice 261, 263, 265
blood oxygen level 36
blood pressure 190, 192–3, 196–8, 200–2
brain waves 36, 39
brakes and braking 8, 107–20, 238
breakdowns 5, 10
buses 8, 45, 69, 121–5

Canada 34–6, 39, 121
carelessness in driving 234, 235–6, 238–42, 244
central nervous system 190, 199, 201
circadian rhythms 25, 34, 156–7, 163, 165, 234 *see also* time of day
co-drivers 9–10, 151–4
  alertness monitoring 225–8
  see also two-up driving
commercial vehicles (CMVs) 33–9, 69, 133
company drivers 15–20, 27–9, 151–4
concurrent set tasks 169, 171, 175, 176–7

countermeasures 26, 97–105
　CMV drivers 33–5, 38–9
　deceleration indicators 107–20
　education programme 60, 121–5
　fatigue 4, 6–9, 62–6, 83, 86, 155, 159, 165
　profile edge lines 127–32
　radio 101–2
　rest breaks 6, 100
　simulators 249–51, 257
　sleep 8–9
　window opening 100–2
critical flicker fusion test 31, 190, 193, 197–9, 201
critical tracking task 133, 137, 139
cyclists 53–7

day of the week 7, 51, 54, 83–4, 98
deceleration indicators 8, 107–20, 238
declarative state of fatigue 156, 158–61, 164–5
descriptions and records 266–9
digit symbol coding 193, 197, 198
distractors 235–6, 246
diurnal cycle *see* time of day
divided attention tasks 144, 190, 193–4, 196–9, 201
double lane change tasks 137–8
DRIVE programme 161
driver behaviour 3–13, 214, 233–47, 261–4, 272
　physiological measures 181–8
driver's hours
　accidents 3, 6, 9, 41–6, 97, 100–4, 133, 219, 226
　alcohol and fatigue 189–90, 194–6, 199–200, 202
　alertness 45–6, 219, 224, 226
　freight 18, 20, 151–4
　livestock transport 23
　methodological research 155–6, 161, 165
　physiology of driver 181, 183–4, 186
　regulations 3, 6–7, 11, 25–6, 29, 33–5, 38, 41–7
　rest breaks 30
　simulator tests 134, 140–2, 146–7
driver's log 41, 133
driving without awareness (DWA) 162–3
drowsiness 7, 12
　alertness monitoring 223–5, 227–8
　driver behaviour 234–5, 238–9, 241, 246
　scale 223
drugs 19, 152, 168, 233–47
　driving and enforcement 67–72
　education programme 8, 121–5
　physiology 181–2, 184, 186

edge lines
　painted 7, 63–5

　profile 65–6, 86, 127–32
electrocardiogram (ECG) 37, 181, 183, 186, 222
electroencephalogram (EEG) 36–7, 139, 181–6, 190, 222–3
electromyogram (EMG) 139, 181, 222
electro-oculogram (EOG) 36–7, 139, 222
emergency tasks 192, 195, 198–9
emotion and driving 181, 244–6
empty lane fallacy 239, 245
environmental conditions 6, 12, 27, 34, 38, 44, 214–15
Europe 4, 127, 219
　ATT 10–11, 161, 164, 186
experienced drivers 3, 11, 14, 62, 152–3, 259–74
　fatigue 158, 164–5, 191
　simulators 135, 253–4
　young drivers 249, 252
eye closures 36, 42–4, 159, 223–4, 226–7
　blinks 220–1

faulty schemata 234, 235–6, 238–9, 245
feedback to drivers 11–12, 160, 163–5, 260, 265
　simulators 12, 253, 256–7
fixed-set tasks 169, 171, 173–4
flashing warning lights 38
freight 5–7, 26, 15–19, 151–4
　specialized 19–20

galvanic skin response (GSR) 181
geographical information systems (GIS) 82–3

health and fitness 3, 7, 11, 30, 157, 164–5, 181
heart rate 31, 36, 39, 221
　alcohol 182–6, 190–1, 193–6, 198–201
heart rate variability 36
　alcohol 182–5, 190–1, 194–6, 199–200
highway hypnosis 76, 82, 162
humidity 6, 37, 38

intelligent vehicle highway systems (IVHS) 186, 219
intersections 51, 54, 62, 185, 252
Iowa Driving Simulator (IDS) 4, 207, 209–16

Jex task 31

lane deviation 11, 42, 159, 162, 183–6
　CMVs 36–7
　countermeasures 8–9
　simulator 133, 138, 144–7, 253
　see also steering
learner drivers 69
livestock transport 6, 21–3
loading 3, 9, 16–17, 26–7, 41, 151
loss of control *see* run-off-road accidents

mental models 266, 268–9, 271, 272
micro-sleeps 9, 76, 162
misattention 234, 235, 238–9, 244, 246
monotony 6–7, 11–12, 25, 183, 199, 201, 219
   accidents 44, 63, 89–90, 93
motion cues 207–13, 216
motorcycles 51, 53–5, 57, 70, 128
multi-task performance 253–5
muscle tone 139, 221

National Advanced Driving Simulator (NADS) 216–17
night-time driving 18, 20, 28
   accidents 7–8, 42–3, 51–4, 88, 92–3
   deceleration indicators 107, 109
   see also time of day
novice drivers 69, 259–60, 269, 272

owner drivers 5–6, 10, 27–9

pedestrians 53–7, 61, 62
   simulators 208, 212, 214
perceptual envelope 259–60, 272
performance operating characteristics (POCs) 255
personality 3, 10, 18, 157, 163–5, 249
physiological functioning 31, 33–4, 36, 39, 181–8
   methodological research 159–60, 164
   simulator tests 134, 139, 142, 145, 147
potentially hazardous stimulus (PHS) 236, 245–6
power supply control problem 269–72
prime movers 15, 19, 20, 151
prior activity 8, 43–4, 46, 91, 93
   see also loading
probationary drivers 69
PROMETHEUS programme 4, 219

random breath testing 67, 69
Rasmussen's model 261–6, 272
reaction time 8, 27, 31, 36, 43
   alcohol 167–78, 189, 194–5, 198–9
   expertise 262, 263
receiver operating characteristics (ROCs) 255
remote areas (WA)
   accidents 7, 51–7, 74, 76–7, 82–6
   driver's hours 44–5
respiration 190–1, 200
rest breaks 30, 38, 42–3, 86, 97, 98, 100–2, 104
   activity during 6, 42–3, 44, 46
reticular activating system (RAS) 190
risk-taking behaviour 234, 239, 241, 246
   expertise 259–60, 272
   young drivers 249–50
road design 7, 12, 216, 260

accidents 63, 87–94
road surface
   accidents 8, 51, 54, 60, 62, 76, 78, 79–82
   simulators 208, 213, 215
road trains 5, 15–17, 22, 151–4
rumble strips 9, 38, 66, 86
run-off-road accidents 9, 62, 76–80, 97, 133
   drugs 70
   road design 88–91
   simulators 136–8, 141, 143, 146–7
rural areas 8, 51–7, 59–62, 73–86, 98

seat belts 57, 87, 97
schedules 3, 10–11, 15–18, 28–9, 31
   CMVs 34, 39
   deliveries 3, 5–6, 18–19
sex of driver 53–5, 135
simulators 133–48, 207–17, 249–57
   alcohol 182
   alertness monitoring 219–29
   driver behaviour 4, 9–10, 12
   young drivers 12, 135, 249, 253–7
single driving *see* solo driving
single-vehicle accidents 7, 42, 60–2, 64–5, 155–6
   age of driver 85–6
   CMVs 133
   drugs 70
   region (WA) 53–4, 57, 73–86
   road design 88–93
sinus arrhythmia 3, 190, 200
skidding 263–4
sleep 25, 98, 100–4
   alcohol 193, 201
   apnoea 158, 164
   in cab 5, 10, 17, 21–2, 151, 153–4
   CMVs 36–7
   countermeasures 8–9
   inadequacy 8, 10–12, 156–8, 162–5, 181
   livestock transport 21–2
   simulator tests 139, 145–6, 219, 222–8
   solo driving 5, 18
   specialized freight 20
   two-up driving 5, 16–18, 151, 153–4
   at the wheel 8, 10, 12, 159, 162–5, 225, 235
      accidents 42–3, 62, 100, 103, 155
solo driving 5, 10, 18, 26–9, 154
speed 9, 31, 42, 97, 159, 246, 252
   accidents 57, 73–4, 87, 89–91, 93
   alcohol 192
   alertness 225, 227–8
   CMVs 37
   deceleration indicators 107
   livestock transport 22–3
   profile edge lines 128–32
   road design 87, 89–91, 93
   road trains 17
   simulator tasks 136–41, 143–7, 216, 252–4

staged driving 6, 26–9, 31
steering 10–11, 27, 31, 43, 159
   alcohol 189, 192, 198
   alertness 45, 222, 225–6
   CMVs 36–7
   driver physiology 182–3, 185–6
   simulator tasks 137–9, 146–7, 208–9, 212, 214
   see also lane deviation
subjective views on fatigue 33, 36–7, 156, 158–61, 164–5
   long distance drivers 26–31

telephoning while driving 185–6, 246
temperature 6, 36, 181
   in cab 34, 37, 38
   thermoregulation 190, 192–3, 197–8, 200–1
thunderstorms 22–3
time of day 11, 25, 154, 156–7, 161, 163
   accidents 6, 42–4, 46, 60–1, 84, 98–104
   age of driver 11
   alertness 219
   driver behaviour 3, 6–7, 9–10
   early morning fatigue 27–8
   personality 18
   physiology of driver 181, 182, 184
   see also circadian rhythms; night-time driving
time-on-task *see* driver's hours
tracking 31, 42, 44, 161, 224, 252
   alcohol 189–90, 192, 194, 196, 198, 201
   expertise 259, 261

traffic volume 107–8
   accidents 63, 76, 82–4, 86
train collision 92–3
truck operator performance system (TOPS) 9, 133–5, 139–40, 142, 144–7
two-up driving 5, 9, 16–18, 21, 26–9, 151–4
   alertness monitoring 225–8
tyres 17, 20, 21, 151, 208

unexpected obstacles 137–8
United States of America 45, 133
   brake lights 110
   CMVs 33–9
   driver's hours 41–2
   edge lines 64
   IVHS 186, 219
   simulator 4, 207, 209–16
urban areas 7, 51–7, 60, 73–86

vagal tone 36–7, 199
varied-set tasks 169–72
ventilation 222
video cameras 37, 39
   alertness monitoring 220–2, 224, 226
vigilance 31, 36, 234–5, 238, 240–4
   alcohol 189, 199–201
   physiology of driver 183–4, 186
visual stimulation 62, 63, 66, 87, 192

weather 22–3, 27, 214–15
weaving *see* lane deviation
workload on driver 181, 183, 185–6